Analysis and Theories in Geosciences

Analysis and Theories in Geosciences

Edited by **Theodore Roa**

New York

Published by Callisto Reference,
106 Park Avenue, Suite 200,
New York, NY 10016, USA
www.callistoreference.com

Analysis and Theories in Geosciences
Edited by Theodore Roa

© 2015 Callisto Reference

International Standard Book Number: 978-1-63239-069-1 (Hardback)

Contents

Permissions

List of Contributors

Preface

This book is a collection of studies and essays regarding geosciences contributed by many of the leading experts in this field. Extracting data from chaotic systems in geosciences has become pretty easy due to fractal analysis. It is being used to deal with a lot of problems in this field. This book presents latest applications of fractal and multifractal assessments in geoscience. It discusses the functioning of the fractal evaluation in climatology and data of cosmic and solar geomagnetic data from experiments. It concludes with special focus on the usage of multi fractal estimation in discovering geophysics. This book will be a significant source of information for researchers and students.

The researches compiled throughout the book are authentic and of high quality, combining several disciplines and from very diverse regions from around the world. Drawing on the contributions of many researchers from diverse countries, the book's objective is to provide the readers with the latest achievements in the area of research. This book will surely be a source of knowledge to all interested and researching the field.

In the end, I would like to express my deep sense of gratitude to all the authors for meeting the set deadlines in completing and submitting their research chapters. I would also like to thank the publisher for the support offered to us throughout the course of the book. Finally, I extend my sincere thanks to my family for being a constant source of inspiration and encouragement.

Editor

Complexity Concepts and Non-Integer Dimensions in Climate and Paleoclimate Research

Reik V. Donner

Additional information is available at the end of the chapter

1. Introduction

The ongoing global climate change has severe effects on the entire biosphere of the Earth. According to the most recent IPCC report [1], it is *very likely* that anthropogenic influences (like the increased discharge of greenhouse gases and a gradually intensifying land-use) are important driving factors of the observed changes in both the mean state and variability of the climate system. However, anthropogenic climate change competes with the natural variability on very different time-scales, ranging from decades up to millions of years, which is known from paleoclimate reconstructions. Consequently, in order to understand the crucial role of man-made influences on the climate system, an overall understanding of the recent system-internal variations is necessary.

The climate during the *Anthropocene*, i.e. the most recent period of time in climate history that is characterized by industrialization and mechanization of the human society, is well recorded in direct instrumental measurements from numerous meteorological stations. In contrast to this, there is no such direct information available on the climate variability before this epoch. Besides enormous efforts regarding climate modeling, conclusions about climate dynamics during time intervals before the age of industrial revolution can only be derived from suitable secondary archives like tree rings, sedimentary sequences, or ice cores. The corresponding paleoclimate proxy data are given in terms of variations of physical, chemical, biological, or sedimentological observables that can be measured in these archives. While classical climate research mainly deals with understanding the functioning of the climate system based on statistical analyses of observational data and sophisticated climate models, paleoclimate studies aim to relate variations of such proxies to those of observables with a direct climatological meaning.

Classical methods of time series analysis used for characterizing climate dynamics often neglect the associated multiplicity of processes and spatio-temporal scales, which result in a very high number of relevant, nonlinearly interacting variables that are necessary for fully describing the past, current, or future state of the climate system. As an alternative, during the last decades concepts for the analysis of complex data have been developed, which are mainly motivated by findings originated within the theory of nonlinear deterministic dynamical systems. Nowadays, a large variety of methods is available for the quantification of the nonlinear dynamics recorded in time series [2,3,4,5,6,7,8,9], including measures of predictability, dynamical complexity, or short- as well as long-term scaling properties, which characterize the dynamical properties of the underlying deterministic attractor. Among others, fractal dimensions and associated measures of structural as well as dynamical complexity are some of the most prominent nonlinear characteristics that have already found wide use for time series analysis in various fields of research.

This chapter reviews and discusses the potentials and problems of fractal dimensions and related concepts when applied to climate and paleoclimate data. Available approaches based on the general idea of characterizing the complexity of nonlinear dynamical systems in terms of dimensionality concepts can be classified according to various criteria. *Firstly*, one can distinguish between methods based on dynamical characteristics estimated directly from a given univariate record and those based on a (low-dimensional) multivariate projection of the system reconstructed from the univariate signal. *Secondly*, one can classify existing concepts related to non-integer or fractal dimensions into self-similarity approaches, complexity measures based on the auto-covariance structure of time series, and complex network approaches. *Finally*, an alternative classification takes into account whether or not the respective approach utilizes information on the temporal order of observations or just their mutual similarity or proximity. In the latter case, one can differentiate between correlative and geometric dimension or complexity measures [10]. Table 1 provides a tentative assignment of the specific approaches that will be further discussed in the following. It shall be noted that this chapter neither gives an exhaustive classification, nor provides a discussion of all existing or possible approaches. In turn, the development of new concepts for complexity and dimensionality analysis of observational data is still an active field of research.

	Methods based on univariate time series	Methods based on multivariate reconstruction
Self-similarity / scaling approaches	*Correlative:* Higuchi estimator for D_0, estimators of the Hurst exponent (R/S analysis, detrended fluctuation analysis, and others)	*Geometric:* fractal dimensions based on box-counting and box-probability, Grassberger-Procaccia estimator for D_2
Approaches based on auto-covariance structure		*Correlative:* LVD dimension density
Complex network approaches	*Correlative:* visibility graph analysis	*Geometric:* recurrence network analysis

Table 1. Classification of some of the most common dimensionality and complexity concepts mainly discussed in this chapter.

In order to illustrate the specific properties of the different approaches discussed in this chapter, the behavior of surface air temperature data is studied. Specifically, the data utilized in the following are validated and homogenized daily mean temperatures for 2342 meteorological stations distributed over Germany (Figure 1) and covering the time period from 1951 top 2006. The raw data have been originally obtained by the German Weather Service for a somewhat lower number of stations before being interpolated and post-processed by the Potsdam Institute for Climate Impact Research for the purpose of validating regional climate simulations ("German baseline scenario"). Before any further analysis, the annual cycle has been removed by means of phase averaging (i.e. subtracting the long-term climatological mean for each calendar day of the year and dividing the residuals by the corresponding empirical standard deviation estimated from the same respective day of all years in the record). This pre-processing step is necessary since the annual cycle gives the main contribution to the intra-annual variability of surface air temperatures in the mid-latitudes and would thus lead to artificially strong correlations on short to intermediate time-scales (i.e. days to weeks) [11]. In addition, since some of the methods to be discussed can exhibit a considerable sensitivity to non-stationarity, linear trends for the residual mean temperatures are estimated by a classical ordinary least-squares approach and subtracted from the de-seasoned record.

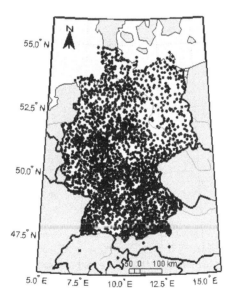

Figure 1. Spatial distribution of the studied surface air temperature records over Germany.

The remainder of this chapter will follow the path from established self-similarity concepts and fractal dimensions (Section 2) over complexity measures based on the auto-covariance structure of time series (Section 3) to modern complex network based approaches of time series analysis (Section 4). Mutual similarities and differences between the individual approaches are addressed. The performance of the different approaches is illustrated using the aforementioned surface air temperature records. Subsequently, the problem of adapting the considered methods to time series with non-uniform (and possibly unknown) sampling as common in paleoclimatology is briefly discussed (Section 5).

2. Self-similarity approach to fractal dimensions

The notion of fractal dimensions has originally emerged in connection with self-similar sets such as Cantor sets or self-similar curves or objects embedded in a metric space [9]. The most classical approach to quantifying the associated scaling properties is counting the number of boxes needed to cover the fractal object under study in dependence on the associated length scale, which behaves like a power-law for fractal systems. More formally, studying the asymptotic behavior of the double-logarithmic dependence between number and size of hypercubes necessary to cover a geometric object with ever decreasing box size defines the box-counting dimension (often also simply called "the" fractal dimension)

$$D_0 = \lim_{l \to 0} \frac{\log N(l)}{\log(1/l)}. \tag{1}$$

Given a trajectory of a complex system in a d-dimensional space that is supposed to correspond to an attractive set, covering the volume captured by this trajectory by hypercubes in the way described above allows estimating the fractal dimension of the associated attractor. More general, considering the probability mass of the individual boxes, p_i, one can easily generalize the concept of box-counting dimensions to so-called Renyi dimensions [12,13]

$$D_q = \lim_{l \to 0} \frac{\log \sum_i p_i^q}{(1-q)\log l}, \tag{2}$$

which give different weights to parts of phase space with high and low density (in fact, the box coverage probabilities p_i serve as naïve estimators of the coarse-grained invariant density $p(x)$ of the dynamical system under study). The special cases $q=0,1,2$ are referred to as the box-counting (or capacity), information, and correlation dimension.

In typical situations, only a univariate time series is given, which can be understood as a low-dimensional projection of the dynamics in the true higher-dimensional phase space. In such cases, it is possible to reconstruct the unobserved components in a topologically equivalent way by means of so-called time-delay embedding [14], i.e. by considering vectors

$$\vec{y}_i = (x_i, x_{i+\tau}, ..., x_{i+(N-1)\tau}), \tag{3}$$

where the unknown parameters N and τ (embedding dimension and delay, respectively) need to be appropriately determined. The basic idea is that the components of the thus reconstructed state vectors are considered to be independent of each other in some feasible sense, thus representing the dynamics of different observables of the studied system. There are some standard approaches for estimating proper values for the two embedding parameters. On the one hand, the delay can be inferred by considering the time after which the serial correlations have vanished (first root of the auto-correlation function) or become statistically insignificant (de-correlation time) – in these cases, the resulting components of the reconstructed state space are considered linearly independent. Alternatively, a measure for general statistical dependence such as mutual information can be considered to estimate the time after which all relevant statistical auto-dependences have vanished [15]. On the other hand, the embedding dimension is traditionally estimated by means of the false nearest-neighbor method, which considers the changes in neighborhood relationships among state vectors if the dimension of the reconstructed phase space is increased by one. Since such changes indicate the presence of projective effects occurring when considering a too low embedding dimension, looking for a value of N for which the neighborhood relationships between the sampled state vectors do not change anymore provides a feasible estimate of the embedding dimension [16]. An alternative approach is considering the so-called singular system analysis (SSA), which allows determining the number of statistically relevant eigenvalues of the correlation matrix of the high-dimensionally embedded original record as an estimate of the true topological dimension of the system under study [17,18].

Having reconstructed the attractor by finding a reasonable approximation of its original phase space as described above, one may proceed with estimating the fractal dimensions by means of box-counting. However, since this approach requires studying the limit of many data, is may become unfeasible for analyzing real-world observational time series of a given length. As alternatives, other approaches have been proposed for estimating some of the generalized fractal dimensions D_q, with the Grassberger-Procaccia algorithm for the correlation dimension [19,20] as the probably most remarkable example. Details on corresponding approaches can be found in any contemporary textbook on nonlinear time series analysis.

A noteworthy alternative to considering fractal dimension estimates based on phase space reconstruction has been introduced by Higuchi [21,22], who studied the behavior of the curve length associated with a univariate time series in dependence on the level of coarse-graining,

$$L_m(k) = \frac{1}{k} \frac{T-1}{\left[\frac{T-m}{k}\right]k} \sum_{i=1}^{\left[\frac{T-m}{k}\right]} \left| x_{m+ik} - x_{m+(i-1)k} \right| \tag{4}$$

(where [.] denotes the integer part), which scales with a characteristic exponent corresponding to the fractal dimension D_0,

$$L(k) = \frac{1}{k} \sum_{m=1}^{k} L_m(k) \propto k^{-D_0}. \tag{5}$$

Figure 2 shows the actual behavior of the thus computed curve length with varying coarse-graining level k (equivalent to the embedding delay in Equation (3)) for the daily mean temperature record from Potsdam. One can see that there are two distinct scaling regimes corresponding to time scales up to about one week and above about ten days. For the shorter time-scales, the slope of the linear fit in the double-logarithmic plot yields values between 1.6 and 1.7, which are of the order of magnitude that is to be expected for low-dimensional chaotic systems with two topological dimensions (note that the drawing of the curve underlying the definition of the curve length $L(k)$ corresponds to a two-dimensional space). In turn, for larger time scales, the slope of the considered function takes values around 2, implying that the dynamics on these time-scales is less structured and resembles a random walk without the distinct presence of an attractive set in phase space with a lower (fractal) dimension. It should be emphasized that the shorter time-scale appears to be coincident with typical durations of large-scale weather regimes, whereas the second range of time-scales exceeds the predictability limit of atmospheric dynamics.

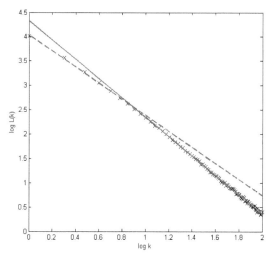

Figure 2. Performance of the Higuchi estimator for the fractal dimension D_0 for the saisonally adjusted and detrended daily mean temperature record from Potsdam. The dashed and solid lines give linear fits to the curve length in double-logarithmic plots based on short (1-7 days) and longer (10-100 days) time scales.

The difference between both scaling regimes becomes even more remarkable when studying the corresponding spatial pattern displayed by all 2342 meteorological stations in Germany (Figure 3). On the shorter time-scales, the fractal dimension is significantly enhanced in the easternmost part of the study area, whereas the same region shows the lowest values of D_0 on the longer time-scales. The presence of two different ranges of time-scales with distinctively different spatial pattern is actually not unique to the fractal dimension, but can also be observed by other complexity measures (see Section 3.5 of this chapter). The probable reason for this finding is the presence of atmospheric processes (related to more marine and continental climates as well as low- and highlands) affecting the different parts of the study area in different ways on short and long time-scales. A more detailed climatological interpretation of this finding is beyond the scope of the present work.

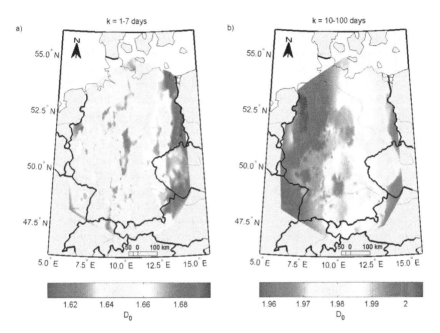

Figure 3. Higuchi estimates of the fractal dimension D_0 for short (1-7 days, panel a) and longer (10-100 days, panel b) time scales.

3. Complexity measures based on serial correlations

As an alternative to concepts based on classical fractal theory, scaling properties based on the linear auto-covariance structure of time series data also contain valuable information. Corresponding approaches utilizing basic methods from multivariate statistics have been referred to as multivariate dimension estimates [11,23,24,25] and provide meaningful characteristics that can be reliably estimated even from rather short time series, which still constitute a fundamental limit for classical fractal dimension analysis.

The original motivation for the introduction of multivariate dimension estimates to climate research has been that the "complete" information about the climate of the past requires considering a set of complementary variables, which form a multivariate time series. The fraction of dynamically relevant observables, which is interpreted as a measure for the average information content of a given variable, can vary itself with time due to the non-stationarity of the climate system. Temporal changes of this information content, i.e. of the effective "dimension" of the record, can therefore serve as an indicator for changes in environmental conditions and the corresponding response of the climate system. Moreover, widely applicable ideas from the theory of nonlinear deterministic processes can be used to adapt this approach to univariate time series. In the following, the mathematical background of the corresponding approach will be detailed.

3.1. Dimensionality reduction of multivariate time series

Quantifying the number of dynamically relevant components in multivariate data sets commonly requires an appropriate statistical decomposition of the data into univariate components with a well-defined variance. In the most common case, these components are required to be orthogonal in the vector space spanned by the original observables, i.e. linearly independent. A corresponding decomposition (typically with the scope of achieving a suitable dimensionality reduction of a given high-dimensional data set) is commonly realized by means of principal component analysis (PCA) [26,27], which is also referred to as empirical orthogonal function (EOF) analysis or Karhunen-Loève decomposition (KLD) depending on the particular scientific context and application. The basic idea beyond this technique is that a proper basis adjusted to the directions of strongest (co-)variation in a multivariate data set can be identified using a principal axis transform of the corresponding correlation matrix. In this case, the associated eigenvectors of the correlation matrix contain weights for linear superpositions of the original observables that result in the largest possible variance. The actual amplitude of this variance is characterized by the associated non-negative eigenvalues.

Technically, consider simultaneous records X_{ij} of different observables $X^{(j)}$ at times t_i combined in a TxN-dimensional data set $\underline{X}=(X_{ij})$ with column vectors representing T successive observations of the same quantity and row vectors containing the simultaneous measurements of N different observables. Here, the columns of $\underline{X}$ may represent different variables measured at the same location or object, or spatially distributed records of the same observable or different variables. The associated correlation matrix is given as the covariance (or scatter) matrix $\underline{S}=\underline{Y}^T\underline{Y}$ where the matrix $\underline{Y}$ is derived from $\underline{X}$ by subtracting the column means from all columns of $\underline{X}$ and then dividing the residual column vectors by their standard deviations. It should be emphasized that column mean and standard deviation represent here estimates of the expectation value and expected standard deviation of the respective observable. The elements of $\underline{S}$ are the linear (Pearson) correlation coefficients between all pairs of variables, which provide reasonable insights into mutual linear interrelationships between the different variables if the observations are normally distributed or the sample size is sufficiently large to neglect the former requirement according to the central limit theorem. By definition, $\underline{S}$ is symmetric and positive semi-definite, i.e. has only non-negative eigenvalues σ^2. Without loss of generality, one may arrange these N eigenvalues in descending order and interpret them as the variances of the principal components of $\underline{X}$ given by the corresponding eigenvectors.

It shall be noted that there are various generalizations of linear PCA, involving decompositions of multivariate data sets into projections onto curved manifolds that take the place of the orthogonal eigenvectors describing the classical linear principal components. Due to the considerably higher computational efforts for identifying these objects in the underlying vector space and correctly attributing the associated component variances, corresponding methods like nonlinear PCA [28], isometric feature mapping (Isomap) [29], or independent component analysis (ICA) [30], to mention only a few examples, will not be further discussed here, but provide possibilities for generalizing the approach detailed in the following.

3.2. KLD dimension density

The idea of utilizing PCA for quantifying the number of dynamically relevant components, i.e. transferring this traditional multivariate statistical technique into a dynamical systems context, is not entirely new. In fact, it has been used as early as in the 1980s for identifying the proper embedding dimension for univariate records based on SSA (see Section 2 of this chapter). Ciliberti and Nicolaenko [31] used PCA for quantifying the number of degrees of freedom in spatially extended systems. Since these degrees of freedom can be directly associated with the fractal dimension or Lyapunov exponents of the underlying dynamical system [32,33,34], it is justified to interpret the number of dynamically relevant components in a multivariate record as a proxy for the effective dimensionality of the corresponding dynamical system.

More formally, Zoldi and Greenside [35,36,37,38] suggested using PCA for determining the number of degrees of freedom in spatially extended systems by considering the minimum number of principal components required to describe a fraction f $(0<f<1)$ of a multivariate record. Let σ_i^2, $i=1,...,N$, again be the non-negative eigenvalues of the associated correlation matrix $\underline{S}$ given in descending order. The aforementioned number of degrees of freedom, which is referred to as the KLD dimension, can then be defined as follows [23]:

$$D_{KLD}(f) = \min\left\{p \left| \sum_{i=1}^{p}\sigma_i^2 \middle/ \sum_{i=1}^{N}\sigma_i^2 \geq f\right.\right\}. \tag{6}$$

For spatially extended chaotic systems, it has been shown that the KLD dimension increases linearly with the system size N, i.e. the number of simultaneously recorded variables [37]. This motivates the study of a normalized measure, the KLD dimension density

$$\delta_{KLD} = D_{KLD} / N, \tag{7}$$

instead of D_{KLD} itself.

3.3. LVD dimension density

While the KLD dimension density can be widely applied for characterizing complex spatio-temporal dynamics based on large data sets (i.e. both N and T are typically large), it reaches its conceptual limits when being applied to multivariate data sets with a small number of simultaneously measured variables (small N), or used for studying non-stationary dynamics in a moving-window framework (small T). On the one hand, small N implies that δ_{KLD} can only have very few distinct values (i.e. multiples of $1/N$), so that small changes in the covariance structure of the considered data set may lead to considerably large changes of the value of this measure. On the other hand, short data sets (small T) imply problems associated with the statistical estimation of correlation coefficients between individual variables (particularly large standard errors and the questionable reliability of the Pearson correlation coefficient as a measure for linear interrelationships in the presence of non-Gaussian distributions). However, both cases can have a considerable relevance in the field of geoscientific data analysis.

As an alternative, Donner and Witt [11,23,24,25] suggested studying the characteristic functional behavior of δ_{KLD} in dependence on the explained variance fraction f. Specifically, if the residual variances decayed exponentially, i.e.

$$V_r(p) = 1 - \sum_{i=1}^{p} \sigma_i^2 \Big/ \sum_{i=1}^{N} \sigma_i^2 = \exp\left(-\frac{p}{N}\Big/\delta\right), \tag{8}$$

the KLD dimension density would scale as

$$\delta_{KLD}(\phi) = -\delta(f)\ln(1-\phi) \quad \text{for} \quad \phi \in [0, f]. \tag{9}$$

in the limit of large N. The resulting coefficient $\delta(f)$ can be understood as characterizing the effective dimensionality of the system. The derived quantity

$$\delta^* = \delta(f) / \log_{10} e \tag{10}$$

(the dependence on f will be omitted for brevity from now on) has been termed the linear variance decay (LVD) dimension density of the underlying data set. Its estimation by means of linear regression according to Equation (9) has been discussed in detail elsewhere [11,25]. It should be mentioned that δ^* does not yet give a properly normalized dimension density with values in the range between 0 and 1, which can already be observed for simple stochastic model systems [23,25]. However, using the limiting cases of identical (lowest possible value δ^*_{min}) and completely uncorrelated (highest possible value δ^*_{max}) component time series, one can derive analytical boundaries and properly renormalize the LVD dimension density to values within the desired range [39] as

$$\delta_{LVD} = \frac{\delta^* - \delta^*_{min}}{\delta^*_{max} - \delta^*_{min}}. \tag{11}$$

It shall be noted that using the LVD dimension density instead of the KLD dimension density solves the problem of discrete values in the limit of small N, but still shares the conceptual limitations with respect to the limit of small T. As another positive feature, δ_{LVD} has a continuous range and a much smaller variability with f than δ_{KLD}. This variability is mainly originates from insufficiencies of the regression model (Equation (8)) and would vanish in case of large N and an exactly exponential decay of the residual variances, which is a situation that is, however, hardly ever met in practice [27].

Possible modifications of the LVD dimension density approach include the consideration of alternative measures of pair-wise statistical association, such as Spearman's rank-order correlation or phase synchronization indices [40,41], which may be of interest in specific applications. Although the formalism described above can applied in exactly the same way to such matrices of similarity measures, the statistical meaning of the corresponding decomposition is not necessarily clear.

3.4. Dimension densities from univariate time series

The previously discussed approach can be easily modified for applications to univariate time series [42]. For this purpose, the correlation matrix $\underline{S}$ of the multivariate record is replaced by the Toeplitz matrix of auto-correlations estimated from a univariate data set. In other words, the PCA commonly utilized for defining the KLD and LVD dimension densities is replaced by an SSA step (i.e. a "PCA for univariate data").

As a particular characteristic of the resulting "univariate dimension densities", it should be emphasized that the obtained results crucially depend on the particularly chosen "embedding" parameters, i.e. the "embedding dimension" N and time delay τ. In case of SSA-based methods, it is common to use an "over-embedding", i.e. a number of time-shifted replications of the original record that is much larger than the actual supposed dimensionality of the studied data. Since serial correlations usually decay with increasing time delay, increasing N beyond a certain value (i.e. adding more and more dimensions to the embedded time series) will not change the number of relevant components in the record anymore. As a consequence, δ_{LVD} asymptotically takes stationary values. In turn, selecting the "embedding delay" τ allows studying the dynamical complexity of time series on various time-scales (i.e. from the minimum temporal resolution of the record to larger scales limited only by the available amount of data). Consequently, δ_{LVD} can change considerably as τ is varied.

3.5. Application: Surface air temperatures

For the purpose of discussing measures of dimensionality based on the auto-covariance structure of an observational record, it is useful to first examine the auto-correlation function itself. As a first example, let us consider again the daily mean temperature record from Potsdam, Germany (Figure 4a). For this time series, the auto-correlations decay within only about 7-10 days to values below 0.2 (Figure 4b). Consequently, using short time delays (below about one week) for embedding temperature records leads to components with considerable mutual correlations. In this case, one can expect a low LVD dimension density, since the information contained in one of the embedded components is already largely determined by the other components. In turn, for larger delays, the embedded components become approximately linearly independent of each other, implying that since correlations are generally weaker, more components need to be taken into account for explaining a given fraction of variance from the multivariate embedded record. Hence, the LVD dimension density should considerably increase with the delay. Indeed, this expectation is confirmed by Figure 4c, which displays a sharp increase of δ_{LVD} with increasing embedding delay τ especially at the scales below one week, whereas there is a saturation for larger delays at values rather close to one.

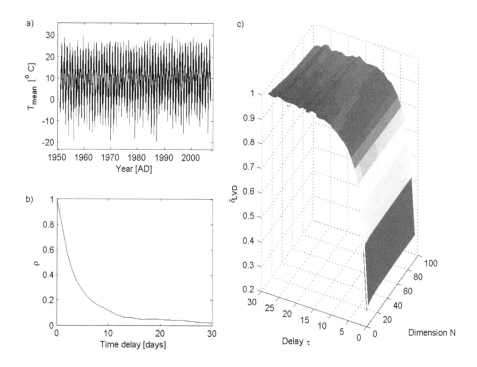

Figure 4. (a) Daily mean surface air temperature record (raw data) from Potsdam, Germany. (b) Corresponding auto-correlation function after de-seasoning. (c) Dependence of the (normalized) LVD dimension density ($f=0.9$) on the embedding parameters dimension and delay.

Another interesting feature can be observed in the behavior of the LVD dimension density with increasing embedding dimension N (Figure 4c). For small delays (i.e. time scales with considerable serial correlations within the observational record), δ_{LVD} increases with increasing N towards an asymptotic value that can be well approximated by estimating this measure for large, but fixed N. In contrast, for large delays, we find a decrease of the estimated LVD dimension density with increasing N without a marked saturation in the considered range of embedding dimensions. A probable reason for this is the insufficiency of the underlying exponential decay model. In fact, the exact functional form of the residual variances for random matrices clearly differs from an exponential behavior, but displays a much more complicated shape [27]. Furthermore, it should be noted that as both delay and embedding dimension increase, the number of available data decreases as $T_{eff}=T-(\tau-1)N$, which can contribute to stronger statistical fluctuations (however, the latter effect is most likely not relevant in the considered example). For intermediate delays, one can thus expect a certain crossover time scale between both types of behavior, which is related to the typical time scale of serial correlations.

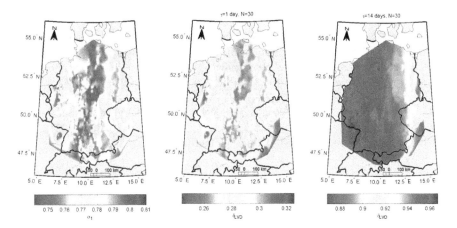

Figure 5. From left to right: AR[1] parameter α_1 and the LVD dimension density δ_{LVD} obtained with $N=30$ and $\tau=1$ day and 14 days, respectively, for the daily mean temperatures over Germany ($f=0.9$).

In order to further support these findings, Figure 5 shows the spatial pattern displayed by the LVD dimension density at all 2342 stations. For larger embedding delays (right panel), the components of the reconstructed multivariate record are in reasonable approximation linearly independent, resulting in high values of the LVD dimension density close to 1 (the limiting case for perfectly uncorrelated records). However, one can observe a marked West/East gradient with high values of δ_{LVD} in the western and central part and much lower values in the eastern part of Germany. Referring to the interpretation of this measure, this finding could indicate that the temporal correlations decay slower in the eastern part that is subject to a more continental climate which typically varies on longer time scales than a marine climate present in the western part of the study area. It should be emphasized that the general spatial pattern closely resembles the behavior of the fractal dimension D_0 (Figure 3b).

In turn, for low embedding delays (1 day), the observed spatial pattern is more complex with more fine-structure, yielding enhanced values (though still indicating considerable correlations) in the eastern and western parts of Germany and lower values in central Germany in a broad band from North to South, as well as in the southeastern part. The qualitative pattern again resembles that of the fractal dimension D_0 (Figure 3a), with the exception that the enhanced values in the eastern part are less well-expressed, whereas the contrasts in the western part are considerably stronger.

In general, both characteristics display similar differences between the behavior on short and longer time scales, which are clearly related to the presence of auto-correlations with a spatially different decay behavior. Regarding the short-term dynamics, this statement is supported by the fact that a qualitatively similar spatial pattern as for the considered dimension estimates (but with opposite trend) can be obtained by coarsely approximating the temperature records by a first-order auto-regressive (AR[1]) process $X_t = \alpha_1 X_{t-1} + \varepsilon_t$, where ε_t is Gaussian white noise (Figure 5, left panel). In fact, for an AR[1] process, the Toeplitz matrix of auto-correlations has a very simple analytical form, $S_{ij} = \alpha_1^{|i-j|}$. Even though there is no closed-form solution for its eigenvalues [43], one can easily show by means of numerical simulations that the resulting LVD dimension density for such processes depends hardly on N, but strongly on the value of the characteristic parameter α_1. Since the latter is related to the time-scale of the associated exponential decay of auto-correlations as $t^* = -1/\log \alpha_1$, low values of α_1 give rise to a fast decay and, hence, high values of the LVD dimension density, whereas the opposite is true for high values close to 1 (see Figure 6). This behavior is in excellent agreement with the theoretical considerations made above.

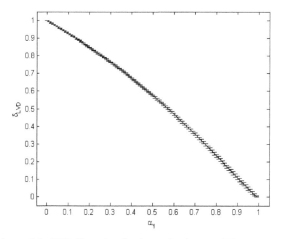

Figure 6. Dependence of the LVD dimension density on the characteristic parameter α_1 of an AR[1] process (f=0.9). The displayed error bars indicate the standard deviations (+/- 1σ) obtained from estimating δ_{LVD} for various embedding dimensions N=2,...,100.

4. Complex network-based approaches

These days, the analysis of network structures is a common task in many fields of science such as telecommunication or sociology, where physical or social interactions (wires, friendships, etc.) can be mathematically described as a graph. When the corresponding connectivity pattern contains a certain number of interacting units (referred to as network vertices or nodes) and is neither completely random nor fully regular (e.g. a chain or lattice), but displays some less obvious type of structure, the resulting system is called a *complex network*. The structural features of such systems can be described using the rich toolbox of quantitative characteristics provided by the so-called complex network theory [44,45,46,47].

Besides the analysis of network structures based on a clearly "visible" substrate (such as infrastructures or communication systems), it has been demonstrated by various authors that complex network approaches can be useful for extracting and understanding the dynamical backbone of systems composed of a large number of dynamically interrelated units or variables, such as financial markets [48], the neuro-physiological activity of different regions of the brain [49], or the functioning of the climate system [50,51,52,53]. In the aforementioned cases, a network structure is identified using suitable measures of statistical association (e.g., linear Pearson correlation or nonlinear mutual information) between records of activity in different areas or of different variables or coupled units. Information on the underlying functional connectivity of the large-scale system is inferred by considering only sufficiently strong interrelationships and studying the set of such connections among the variety of subsystems.

In parallel to the development of complex network methods as a complementary tool for multivariate time series analysis, a variety of different approaches has been suggested for studying single univariate time series from a network perspective [54]. Existing approaches include methods based on transition probabilities after coarse-graining the time series' range or the associated reconstructed phase space [55], convexity relationships between different observations in a record [56], or certain notions of spatial proximity between different parts of a trajectory [57,58,59,60,61,62,63], to mention only the most prominent existing concepts in this evolving area of research (for a more detailed recent review, see [54]). For two of these approaches, the so-called visibility graphs and recurrence networks discussed below, it has been shown that some of the resulting network properties can be related to the concept of fractal dimensions or, more general, scaling analysis. In the following, the corresponding recent findings are summarized.

4.1. Visibility graph analysis

Visibility graphs have been originally introduced as a versatile tool for studying visibility relationships between objects in architecture or robot motion planning [64,65,66,67]. Lacasa and co-workers [55] suggested transferring this idea to the analysis of time series from complex systems, where local maxima and minima of the considered observable play the role of hills and valleys in a one-dimensional landscape. Specifically, in a visibility graph constructed from a univariate time series, the individual observations are taken as network vertices, and edges are established between pairs of vertices $x_i=x(t_i)$ and $x_j=x(t_j)$ that are "mutually visible" from each other, i.e. where for all $x_k=x(t_k)$ with $t_i<t_k<t_j$ the following local convexity condition applies:

$$x_k < x_j + \left(x_i - x_j\right)\frac{t_j - t_k}{t_j - t_i}. \tag{12}$$

When describing the connectivity of this network in the most common way in terms of the binary adjacency matrix A_{ij} (here, $A_{ij}=1$ implies that there exists an edge between vertices i and j), the latter can be consequently expressed as follows:

$$A_{ij}^{VG} = A_{ji}^{VG} = \prod_{k=i+1}^{j-1} \Theta\left(x_j + \left(x_i - x_j\right)\frac{t_j - t_k}{t_j - t_i} - x_k \right), \tag{13}$$

where Θ denotes the Heaviside function defined in the usual way.

As a simplification of the standard visibility graph algorithm, it can be useful considering the so-called horizontal visibility graph, in which the connectivity is defined according to the horizontal visibility between individual vertices, i.e. there is an edge (i,j) between two observations x_i and x_j if for all k with $t_i < t_k < t_j$, $x_k < min(x_i, x_j)$. Consequently, the associated adjacency matrix reads

$$A_{ij}^{HVG} = A_{ji}^{HVG} = \prod_{k=i+1}^{j-1} \Theta\left(x_i - x_k\right)\Theta\left(x_j - x_k\right). \tag{14}$$

In other words, the horizontal visibility graph encodes the distribution of local maxima in a time series (i.e. short-term record-breaking events). Due to its simpler analytical form, it has the advantage that certain basic network properties can be more easily evaluated analytically than for the standard visibility graph.

As a particularly remarkable result, it has been demonstrated both analytically and numerically that for fractal as well as multifractal processes, the degree distributions $p(k)$ of visibility graphs, i.e. the probabilities of finding vertices with a given number of connections (degree)

$$k_i = \sum_j A_{ij}, \tag{15}$$

exhibit a power-law (commonly called "scale-free property" in complex network theory) with a characteristic scaling exponent that is directly related to the associated Hurst exponent H [68,69]. Moreover, it can be shown that for a wide class of such processes, the Hurst exponent is itself related with the fractal dimension D_0 as $D_0 = 2 - H$, however, this relationship is not universal [70]. In this spirit, the scaling exponent obtained from visibility graphs can be considered as an alternative estimate of the fractal dimension. In turn, besides the validity of the aforementioned relationship between Hurst exponent and fractal dimension for the specific data set under study, the possible improvements with respect to computational efforts, required data volume and related issues still need to be systematically compared with those of existing estimators of the Hurst exponent.

In addition to the potentially ambiguous interdependence between Hurst exponent and fractal dimension, using visibility graph approaches for the purpose of estimating fractal dimensions from geoscientific time series may be affected by a further problem. Towards the ends of a time series, there is a systematic tendency to underestimate the actual degree of vertices just due to a lower number of possible neighbors [71]. While this feature will have negligible influence for long time series, it may considerably contribute to a bias in the degree distribution estimated from small data sets common to many geoscientific problems. In turn, a potential advantage of visibility graphs is that they do not require uniform sampling in time, which makes them applicable to typically problematic types of data such as paleoclimate records [71] or even marked point process data such as earthquake catalogues [72].

4.2. Recurrence network analysis

Recurrence networks are another well-studied approach for transforming time series into an associated complex network representation [60,61,62,63]. In contrast to visibility graphs, the basic idea is reconstructing the spatial structure of the attractor underlying the observed dynamics in the corresponding phase space. That is, given a univariate record the dynamically relevant variables need to be reconstructed by means of time-delay embedding first if necessary. Consequently, the first step of recurrence network analysis consists of identifying the appropriate embedding parameters by means of the corresponding standard techniques. Having determined these parameters, time-delay embedding is performed. For the resulting multivariate time series, the mutual distances between all resulting sampled state vectors (measured in terms of a suitable norm in phase space, such as Manhattan, Euclidean, or maximum norm) are compared with a predefined global threshold value ε. Interpreting the state vectors as vertices of a recurrence network, only such pairs of vertices are connected that are mutually closer than this threshold, resulting in the following definition of the adjacency matrix:

$$A_{ij}^{RN} = \Theta\left(\varepsilon - \left\|x_i - x_j\right\|\right) - \delta_{ij},$$ (16)

where δ_{ij} denotes Kronecker's delta defined in the usual way. To put it differently, in a recurrence network only neighboring state vectors taken from the sampled trajectory of the system under study are connected. In this spirit, the recurrence network forms the structural backbone of the associated dynamical system. Moreover, since no information on temporal relationships enters the construction of the recurrence network, its study corresponds to a completely geometric analysis method.

The structural properties of recurrence networks have already been intensively studied. Relating to the degree distributions, it has been demonstrated analytically as well as numerically that the presence of a power-law-shaped singularity of the invariant density $p(x)$ of the studied dynamical system is a necessary condition for the emergence of scale-free degree distributions, the scaling exponent of which is, however, not necessarily associated with the system's fractal dimension, but with the characteristic behavior of the invariant density near its singularity [73]. More generally, recurrence networks are a special case of random geometric graphs aka spatial networks, where the network vertices have a distinct position in some metric space and the connectivity pattern is exclusively determined by the spatial density of vertices and their mutual distances [74]. The latter observation allows calculating expectation values of most relevant complex network characteristics given that the invariant density is exactly known or can at least be well approximated numerically [75]. Specifically, the transitivity properties of recurrence networks on both local and global scale can be computed analytically for some simple special cases [75]. A detailed inspection of these properties demonstrates that the global recurrence network transitivity

$$T = \frac{\sum_{i,j,k} A_{ij} A_{ik} A_{jk}}{\sum_{i,j,k} A_{ij} A_{ik}}$$ (17)

can be considered as an alternative measure for the effective dimensionality of the system under study [76]. In contrast to established notions of fractal dimensions, the estimation of the associated transitivity dimension

$$D_T(\varepsilon) = \log T(\varepsilon) / \log(3/4) \qquad (18)$$

does *not* require considering any scaling properties of some statistical characteristics. The definition in Equation (18) is motivated by the fact that at least using the maximum norm, for random geometric graphs in integer dimensions d the expectation value of the network transitivity scales as $(3/4)^d$ [74,76]. However, it should be noted that the proper evaluation of the transitivity dimension is challenged by the fact that it alternates between two asymptotic values, referred to as upper and lower transitivity dimension, as the recurrence threshold ε is varied [76].

According to the aforementioned interpretation of the transitivity properties, it has been found that the associated local clustering coefficient providing a measure of transitivity on the level of an individual vertex is a sensitive tracer of dynamically invariant objects like supertrack functions or unstable periodic orbits [54,60,61,76]. In turn, global clustering coefficient (i.e. the arithmetic mean value of the local clustering coefficients of all vertices in the recurrence network) and network transitivity track changes in the dynamical complexity of a system under study that are related with bifurcations [10,61,77,78] or subtle changes in the dynamics not necessarily captured by traditional methods of time series analysis [10,79]. In a similar fashion, some other network measures based on the concept of shortest paths on the graph can be utilized for similar purposes. In summary, it has to be underlined that the recurrence network concept has already demonstrated its great potential for studying geoscientific time series, however, this potential has not yet been fully and systematically explored for different fields of geosciences.

5. Complexity and dimensionality analysis in paleoclimatology

Unlike for data obtained from meteorological observatories or climate models, the appropriate statistical analysis of paleoclimate proxy data is a challenging task. Particularly, a variety of technical problems arise due to the specific properties of this kind of data [24].

Firstly, paleoclimate data sets are usually very noisy due to significant measurement uncertainties, high-frequency variations, secondary (non-climatic) effects and the aggregation of the measurements over certain, not necessarily exactly known time intervals.

Secondly, in Earth history environmental conditions have changed both continuously and abruptly, on very long time-scales as well as on a set of different "natural" frequencies the influence of which has changed with time. Especially during the last million years, there has been a sequence of time intervals with cold (glacial) and moderate (interglacial) global climate conditions, which can be interpreted as disjoint states of the global climate system. Even more, these two types of states have alternated in a way that displays some complex regularity, i.e., the timing of the (rather abrupt) transitions between subsequent states (glacial terminations and inceptions) has been controlled by dominating frequencies of variations in the Earth's orbital parameters [80], which is commonly referred to as Milankovich variability. As a consequence of these multiple transitions, paleoclimate time series are intrinsically non-stationary with respect to variability on a variety of different time scales.

Finally, in the case of sedimentary and ice core sequences as the most common types of proxy records, the core depth has to be translated into an age value with usually rather coarse and uncertain age estimates [81,82]. Since the rate of material accumulation has typically varied with time as well, an equidistant sampling along the sequence does usually not imply a uniform spacing of observations along the time axis. Both unequal spacing of measurements and uncertainties in both timing and value pose additional challenges to any kind of time series analysis approach applied to paleoclimate data.

5.1. Analysis of time series with non-uniform sampling

As stated above, non-uniform sampling is an inherent feature of most paleoclimate records. Hence, the appropriate statistical analysis of such records requires a careful specific treatment, since standard estimators of even classical and conceptually simple linear characteristics are not directly applicable (or at least do not perform well) in case of unequally spaced time series data. Consequently, in the last decades there has been an increasing interest in developing alternative estimators that generalize the established ones in a sophisticated way.

Traditionally, many approaches for analyzing paleoclimate time series have implicitly assumed a linear-stochastic behavior of the underlying system, i.e. that the major features of the records can be described by "classical" statistical approaches like correlation or spectral analysis [83,84]. In particular, novel estimators for both time and frequency domain characteristics have been developed which do not require a uniform sampling [85,86,87,88,89]. In turn, many recent studies in the field of paleoclimatology, including such dealing with sophisticated statistical methods [84,90], have typically made use of interpolation to uniform spacing. It has to be underlined that this strategy, however, disregards important conceptual problems such as the appearance of spurious correlations in interpolated paleoclimate data [86] or the presence of time-scale uncertainty. At least the former problem can be solved by using improved more generally applicable estimators, whereas the impact of time-scale uncertainty can be estimated using resampled (Monte Carlo) age models and distributions of statistical properties estimated from ensembles of perturbed age models consistent with the original one [71].

Moreover, classical statistical methods such as correlation or spectral analysis are typically based on the assumption that the observed system is in an equilibrium state, which is reflected by the stationarity of the observed time series. However, this stationarity condition is usually violated in the case of paleoclimate data due to the variable external forcing (solar irradiation) and multiple feedback mechanisms in the climate system that drive the system towards the edge of instability. Hence, more sophisticated methods are required allowing to cope with non-stationary data as well [91]. One prominent example for such approaches is wavelet analysis [92,93,94], which allows a time-dependent characterization of the variability of a time series on different time scales. As for the classical methods of correlation and spectral analysis designed for stationary data, estimators of the wavelet spectrogram are meanwhile also available for unevenly sampled data, for example, in terms of the weighted wavelet Z transform [95,96,97,98,99,100], gapped wavelets [101,102], or a generalized multi-resolution analysis [103,104]. Similar as for classical spectral analysis, such wavelet-based methods often exhibit scaling laws associated with fractal or multifractal properties of the system under study.

5.2. Fractal dimensions and complexity concepts in paleoclimate studies

The question of whether climate can be approximately described as a low-dimensional chaotic system has stimulated a considerable amount of research in the last three decades. Notably, much of the corresponding work has been related with the study of paleoclimate records. As a prominent example, the seminal paper "Is there a climatic attractor?" by Nicolis and Nicolis [105] considered the estimation of the correlation dimension D_2 of the oxygen isotope record from an equatorial Pacific deep-sea sediment core. A direct follow-up [106] presented a thorough re-analysis of the same record utilizing the information dimension D_1. Both manuscripts started an intensive debate on the conceptual as well as analytical limits of fractal dimension estimates for paleoclimate time series. Grassberger [107] analyzed different data sets and could not find any clear indication for low-dimensional chaos. This absence of positive results has been at least partially triggered by the problematic properties of paleoclimate records, particularly the relatively small amount of data and their non-uniform sampling resulting in the need for interpolating the observational time series. Grassberger's results were confirmed by Maasch [108] who analyzed 14 late Pleistocene oxygen isotope records and concluded that "the dimension cannot be measured accurately enough to determine whether or not it is fractal". Fluegeman and Snow [109] used R/S analysis to estimate the fractal dimension D_0 of a marine sediment record via the associated Hurst exponent H, whereas Schulz *et al.* [110] used the Higuchi estimator for a similar purpose. Mudelsee and Stattegger [111,112,113] estimated the correlation dimension of various oxygen isotope records using the classical Grassberger-Procaccia algorithm.

Due to the inherent properties of paleoclimate data, estimating fractal dimensions and related complexity measures is a challenging task. Instead of using the classical fractal dimension concepts, in the last years it has therefore been suggested to consider alternative methods that allow quantifying the dimensionality of such records. Donner and Witt [11,23,24] utilized the multivariate version of the LVD dimension density (see Section 3.3) for studying long-term dynamical changes in the Antarctic offshore sediment decomposition associated with the establishment of significant oceanic currents across the Drake passage at the Oligocene-Miocene boundary. In a similar way, Donges *et al.* used recurrence network analysis for sliding windows in time for identifying time intervals of subtle large-scale changes in the terrigenous dust flux dynamics off North Africa during the last 5 million years [10,79]. These few examples underline the potentials of the corresponding approaches for a nonlinear characterization of paleoclimate records.

5.3. Perspectives and challenges

Both classical as well as novel approaches to characterizing the dimensionality of paleoclimate records still face considerable methodological challenges. While established methods typically rely on the availability of long time series, this requirement can be relaxed when using correlation- or network-based approaches, which are in principle suited for studying nonlinear properties in a running windows framework and thus characterizing the time-varying complexity of environmental conditions encoded in the respective proxy variable under study. However, some methodological challenges persist, which have been widely neglected in the recent literature.

Most prominently, the exact timing of observations is of paramount importance for essentially all methods of time series analysis. In the presence of time-scale uncertainty inherent to most paleoclimate records, this information is missing and can only be incorporated into the statistical analysis by explicitly accounting for the multiplicity of age-depth relationships consistent with the set of available dating points. The latter can be achieved by performing the same analysis for a large set of perturbed age models generated by Monte Carlo-type algorithms, or by incorporating the associated time-scale uncertainty by means of Bayesian methods. However, an analytical theory based on the Bayesian framework can hardly be achieved for all possible methods of time series analysis, so that it is most likely that one has to rely on numerical approximations.

Even when neglecting time-scale uncertainty, the non-uniformity of sampled data points in time typically persists. Among all methods discussed in this chapter, only the visibility graph approach is able by construction to directly work with arbitrarily sampled data. However, this method is faced with the conceptual problem of how to treat values between two successive observations that have not been observed for whatever reason. Donner and Donges [71] argued that simply neglecting such "missing values" may account for a considerable amount of error in all relevant network measures, so that the meaningful interpretability of the obtained results could become questionable.

For the other mentioned approaches, time-delay embedding is a typical preparatory step for all analyses. Since interpolation can result in spurious correlations [86] or at least ambiguous results depending on the specific procedure, alternatives need to be considered for circumventing this problem. In the case of uni- and multivariate LVD dimension density, it is possible to directly utilize alternative estimators of the correlation function, e.g. based on suitable kernel estimates [86], for obtaining the correlation matrix of the record under study. For methods requiring attractor reconstruction (e.g. the Grassberger-Procaccia algorithm for the correlation dimension or recurrence network analysis), there are prospective approaches for alternative embedding techniques, e.g. based on Legendre coordinates [114], that shall be further investigated in future work.

6. Conclusions

Since the introduction of fractal theory to the study of nonlinear dynamical systems, this field has continuously increased its importance. Besides providing a unified view on scaling properties of various statistical characteristics in space or time that can be found in many complex systems, fractal dimensions have demonstrated their great potential to quantitatively distinguish between time series obtained under different conditions or at different locations, thus contributing to a classification of behaviors based on nonlinear dynamical properties. However, as it has been demonstrated both empirically and numerically, established concepts of fractal dimensions reach their fundamental limits when being applied to relatively short and noisy geoscientific time series, e.g. climate records. As potential alternatives providing measures with comparable meaning, but different conceptual foundations, two promising approaches based on the evaluation of serial correlations and complex network theory have been discussed. Although both concepts still need to systematically prove their capabilities and require further methodological improvements as highlighted in this chapter, they constitute promising new research avenues for future problems in climate change research, other fields of geosciences, and even complex systems sciences in general.

Author details

Reik V. Donner
Research Domain IV – Transdisciplinary Concepts and Methods,
Potsdam Institute for Climate Impact Research, Potsdam, Germany

Acknowledgement

This work has been supported by the Leibniz Society (project ECONS).

7. References

[1] Intergovernmental Panel on Climate Change (IPCC). Climate Change 2007: The Physical Science Basis. Cambridge: Cambridge University Press; 2007.

[2] Eckmann J-P, Ruelle D. Ergodic theory of chaos and strange attractors. Reviews of Modern Physics 1985;57(3): 617-56, doi:10.1103/RevModPhys.57.617

[3] Abarbanel, HDI, Brown R, Sidorowich JJ, Tsimring LS. The analysis of observed chaotic data in physical systems. Reviews of Modern Physics 1993;65(4): 1331-92, doi: 10.1103/RevModPhys.65.1331

[4] Tong, H. Nonlinear Time Series and Chaos. Singapore: World Scientific; 1993.

[5] Abarbanel, HDI. Analysis of Observed Chaotic Data. New York: Springer; 1996.

[6] Kantz H, Schreiber T. Nonlinear Time Series Analysis. Cambridge: Cambridge University Press; 1997.

[7] Diks C. Nonlinear Time Series Analysis: Methods and Applications. Singapore: World Scientific; 1999.

[8] Galka A. Topics in Nonlinear Time Series Analysis – With Implications for EEG Analysis. Singapore: World Scientific; 2000.

[9] Sprott JC. Chaos and Time Series Analysis. Oxford: Oxford University Press; 2003.

[10] Donges JF, Donner RV, Rehfeld K, Marwan N, Trauth MH, Kurths J. Identification of dynamical transitions in marine palaeoclimate records by recurrence network analysis. Nonlinear Processes in Geophysics 2011;18(5): 545-62, doi:10.5194/npg-18-545-2011.

[11] Donner R, Sakamoto T, Tanizuka N. Complexity of Spatio-Temporal Correlations in apanese Air Temperature Records. In: Donner RV, Barbosa SM. (eds.) Nonlinear Time Series Analysis in the Geosciences - Applications in Climatology, Geodynamics, and Solar-Terrestrial Physics. Berlin: Springer; 2008. p125-154, doi:10.1007/978-3-540-78938-3_7.

[12] Grassberger P. Generalized dimensions of strange attractors, Physics Letters A 1983;97(6): 227-230, doi:10.1016/0375-9601(83)90753-3.

[13] Hentschel HE, Procaccia I. The infinite number of generalized dimensions of fractals and strange attractors, Physica D 1983;8(3): 435-444, doi:10.1016/0167-2789(83)90235-X.

[14] Packard NH, Crutchfield JP, Farmer JD, Shaw RS. Geometry from a time series, Physical Review Letters 1980;45(9): 712-716, doi:10.1103/PhysRevLett.45.712.

[15] Fraser AM, Swinney HL. Independent coordinates for strange attractors from mutual nformation, Physical Review A 1986;33(2): 1134-1140, doi:10.1103/PhysRevA.33.1134.

[16] Kennel MB, Brown R, Abarbanel HDI. Determining embedding dimension for phase-space reconstruction using a geometrical construction, Physical Review A 1992;45(6): 3403-3411, doi:10.1103/PhysRevA.45.3403.

[17] Broomhead DS, King GP. Extracting qualitative dynamics from experimental data, Physica D 1986;20(2-3): 217-236, doi:10.1016/0167-2789(86)90031-X.

[18] Vautard R, Ghil M. Singular spectrum analysis in nonlinear dynamics, with applications o paleoclimatic time series, Physica D 1989;35(3): 395-424, doi:10.1016/0167-2789(89)90077-8.

[19] Grassberger P, Procaccia I. Characterization of strange attractors, Physical Review Letters 1983;50(5): 346-349, doi:10.1103/PhysRevLett.50.346.

[20] Grassberger P, Procaccia I. Measuring the strangeness of strange attractors, Physica D 1983;9(1-2): 189-208, doi:10.1016/0167-2789(83)90298-1.

[21] Higuchi T. Approach to an irregular time series on the basis of the fractal theory, Physica D 1988;31(2): 277-283, doi:10.1016/0167-2789(88)90081-4.

[22] Higuchi T. Relationship between the fractal dimension and the power law index for a ime series: A numerical investigation, Physica D 1990;46(2): 254-264, doi:10.1016/0167-2789(90)90039-R.

[23] Donner R, Witt A. Characterisation of long-term climate change by dimension estimates of multivariate palaeoclimate proxy data. Nonlinear Processes in Geophysics 2006;13(5): 485-97, doi: 10.5194/npg-13-485-2006.

[24] Donner R, Witt A. Temporary dimensions of multivariate data from paleoclimate records - A novel measure for dynamic characterization of long-term climate change. nternational Journal of Bifurcation and Chaos 2007;17(10): 3685-9, doi:10.1142/S0218127407019573.

[25] Donner R. Advanced Methods for Analysing and Modelling of Multivariate Palaeoclimatic Time Series. PhD thesis. University of Potsdam; 2005.

[26] Joliffe JT. Principal Component Analysis. New York: Springer; 1986.

[27] Preisendorfer RW. Principal Component Analysis in Meteorology and Oceanography. Amsterdam: Elsevier; 1988.

[28] Kramer MA. Nonlinear principal component analysis using autoassociative neural networks. American Institute for Chemical Engineering Journal 1991;37(2): 233-43, doi: 10.1002/aic.690370209.

[29] Tenenbaum J, de Silva V, Langford JC. A global geometric framework for nonlinear diomensionality reduction. Science 2000;290(5500): 2319-23, doi: 10.1126/science.290.5500.2319.

[30] Hyvärinen A, Karhunen J, Oja E. Independent Component Analysis. New York: Wiley; 2001.

[31] Ciliberti S, Nicolaenko B. Estimating the number of degrees of freedom in spatially extended systems. Europhysics Letters 1991;14(4): 303-8, doi: 10.1209/0295-5075/ 14/4/003.

[32] Pomeau Y. Measurement of the information density in turbulence. Comptes Rendus de 'Academie des Sciences. Series II 1985;300(7): 239-41.

[33] Kaneko K. Spatiotemporal chaos in one-dimensional and two-dimensional coupled map attices. Physica D 1989;37(1-3): 60-82, doi: 10.1016/0167-2789(89)90117-6.

[34] Mayer-Kress G, Kaneko K. Spatiotemporal chaos and noise. Journal of Statistical Physics 1989;54(5-6): 1489-508, doi:10.1007/BF01044730.

[35] Zoldi SM, Greenside HM. Karhunen-Loeve decomposition of extensive chaos. Physical Review Letters 1997;78(9): 1687-90, doi: 10.1103/PhysRevLett.78.1687.

[36] Zoldi SM, Liu J, Bajaj KMS, Greenside HM, Ahlers G. Extensive scaling and nonuniformity of the Karhunen-Loeve decomposition for the spiral-defect chaos state. Physical Review E 1998;58(6): 6903-6. doi: 10.1103/PhysRevE.58.R6903.

[37] Meixner M, Zoldi SM, Bose S, Schöll E. Karhunen-Loeve local characterization of spatiotemporal chaos in a reaction-diffusion system. Physical Review E 2000;61(2): 1382-85, doi: 10.1103/PhysRevE.61.1382.

[38] Varela H, Beta C, Bonnefort A, Krischer K. Transitions in electrochemical turbulence. Physical Review Letters 2005;94(17): 174104, doi: 10.1103/PhysRevLett.94.174104.

[39] Xie X, Zhao X, Fang Y, Cao Z, He G. Normalized linear variance decay dimension density and its application of dynamical complexity detection in physiological (fMRI) ime series. Physics Letters A 2011;375(17): 1789-95, doi: 10.1016/j.physleta.2011.03.003.

[40] Donner R. Spatial Correlations of River Runoffs in a Catchment. In: Kropp J, Schellnhuber HJ. (eds.) In Extremis - Disruptive Events and Trends in Climate and Hydrology. Berlin: Springer; 2011. p286-313. doi:10.1007/978-3-642-14863-7_14.

[41] Donner R. Multivariate analysis of spatially heterogeneous phase synchronisation in complex systems: Application to self-organised control of material flows in networks. European Physical Journal B 2008;63(3): 349-61, doi:10.1140/epjb/e2008-00151-8.

[42] Toonen C, Lappe D, Donner RV, Scholz-Reiter B. Impact of Machine-Driven Capacity Constellations on Performance and Dynamics of Job-Shop Systems. In: El Maraghy HA. ed.) Enabling Manufacturing Competitiveness and Economic Sustainability. Berlin: Springer; 2011. p611-6, doi:0.1007/978-3-642-23860-4_100.

[43] Grenander U, Szegö G. Toeplitz Forms and Their Applications. Berkeley: University of California Press; 1958.

[44] Albert R, Barabasi AL. Statistical mechanics of complex networks, Reviews of Modern Physics 2002;74(1): 47-97, doi:10.1103/RevModPhys.74.47.

[45] Newman MEJ. The structure and function of complex networks, SIAM Review 2003;45(2): 167-256, doi:10.1137/S003614450342480.

[46] Boccaletti S, Latora V, Moreno Y, Chavez M, Hwang DU. Complex networks: Structure and dynamics, Physics Reports 2006;424(4-5): 175-308, doi:10.106/j.physrep.2005.10.009.

[47] Costa LF, Rodrigues FA, Travieso G, Villas Boas PR. Characterization of complex networks: A survey of measurements, Advances in Physics 2007;56(1): 167-242, doi:10.1080/00018730601170527.

[48] Kenett DY, Tumminello M, Madi A, Gur-Gershgoren G, Mantegna RN, Ben-Jacob E. Dominating clasp of the financial sector revealed by partial correlation analysis of the stock market, PLoS ONE 2010;5(12): e15032, doi:10.1371/journal.pone.0015032.

[49] Zhou C, Zemanova L, Zamora G, Hilgetag CC, Kurths J. Hierarchical organization unveiled by functional connectivity in complex brain networks, Physical Review Letters 2006;97(23): 238103, doi:10.1103/PhysRevLett.97.238103.

[50] Tsonis AA, Roebber PJ. The architecture of the climate network, Physica A 2004;333: 497-504, doi:10.1016/j.physa.2003.10.045.

[51] Yamasaki K, Gozolchiani A, Havlin S. Climate networks around the globe are significantly affected by El Nino, Physical Review Letters 2008;100(22): 228501, doi:10.1103/PhysRevLett.100.228501.

[52] Donges JF, Zou Y, Marwan N, Kurths J. The backbone of the climate network, Europhysics Letters 2009;87(4): 48007, doi:10.1209/0295-5075/87/48007.

[53] Donges JF, Schultz HCH, Marwan N, Zou Y, Kurths J. Investigating the topology of nteracting networks: Theory and application to coupled climate subnetworks, European Physical Journal B 2011;84(4): 635-651, doi:10.1140/epjb/e2011-10975-8.

[54] Donner RV, Small M, Donges JF, Marwan N, Zou Y, Xiang R, Kurths J. Recurrence-based time series analysis by means of complex network methods, International Journal of Bifurcation and Chaos 2011;21(4): 1019-1046, doi:10.1142/S0218127411029021.

[55] Nicolis G, Garcia Cantu A, Nicolis C. Dynamical aspects of interaction networks, nternational Journal of Bifurcation and Chaos 2005;15(11): 3467-3480, doi:10.1142/S0218127405014167.

[56] Lacasa L, Luque B, Ballesteros F, Luque J, Nuno JC. From time series to complex networks: The visibility graph, Proceedings of the National Academy of Sciences 2008;105(13): 4972-4975, doi:10.1073/pnas.0709247105.

[57] Zhang J, Small M. Complex network from pseudoperiodic time series: Topology versus dynamics, Physical Review Letters 2006;96(23): 238701, doi: 10.1103/PhysRevLett.96.238701.

[58] Yang Y, Yang H. Complex network-based time series analysis, Physica A 2008;387(5-6): 1381-1386, doi:10.1016/j.physa.2007.10.055.

[59] Xu X, Zhang J, Small M. Superfamily phenomena and motifs of networks induced from ime series, Proceedings of the National Academy of Sciences 2008;105(50): 19601-19605, doi:10.1073/pnas.0806082105.

[60] Marwan N, Donges JF, Zou Y, Donner RV, Kurths J. Complex network approach for recurrence analysis of time series, Physics Letters A 2009;373(46): 4246-4254, doi:10.1016/j.physleta.2009.09.042.

[61] Donner RV, Zou Y, Donges JF, Marwan N, Kurths J. Recurrence networks – a novel paradigm for nonlinear time series analysis, New Journal of Physics 2010;12(3): 033025, doi:10.1088/1367-2630/12/3/033025.

[62] Donner RV, Zou Y, Donges JF, Marwan N, Kurths J. Ambiguities in recurrence-based complex network representations of time series, Physical Review E 2010;81(1): 015101(R), doi:10.1103/PhysRevE.81.015101.

[63] Gao Z, Jin N. Flow-pattern identification and nonlinear dynamics of gas-liquid two-phase flow in complex networks, Physical Review E 2009;79(6): 066303, doi:10.1103/PhysRevE.79.066303.

[64] Lozano-Perez T, Wesley MA. An algorithm for planning collision-free paths among polyhedral obstacles, Communications of the ACM 1979;22(10): 560-570, doi:10.1145/359156.359164.

[65] de Floriani L, Marzano P, Puppo E. Line-of-sight communication on terrain models, nternational Journal of Geographical Information Systems 1994;8(4): 329-342, doi:10.1080/02693799408902004.

[66] Nagy G. Terrain visibility, Computers & Graphics 1994;18(6): 763-773, doi:10.1016/0097-8493(94)90002-7.

[67] Turner A, Doxa M, O'Sullivan D, Penn A. From isovists to visibility graphs: A methodology for the analysis of architectural space, Environment and Planning B 2001;28(1): 103-121, doi:10.1068/b2684.

[68] Lacasa L, Luque B, Luque J, Nuno JC. The visibility graph: A new method for estimating the Hurst exponent of fractional Brownian motion, Europhysics Letters 2009;86(3): 30001, doi:10.1209/0295-5075/86/30001.

[69] Ni XH, Jiang ZQ, Zhou WX. Degree distributions of the visibility graphs mapped from ractional Brownian motions and multifractal random walks, Physics Letters A 2009;373(42): 3822-3826, doi:10.1016/j.physleta.2009.08.041.

[70] Gneiting T, Schlather M. Stochastic models that separate fractal dimension and the Hurst effect, SIAM Review 2004;46(2): 269-282, doi:10.1137/S0036144501394387.

[71] Donner RV, Donges JF. Visibility graph analysis of geophysical time series: Potentials and possible pitfalls, Acta Geophysica 2012;60(3): 589-623, doi:10.2478/s11600-012-0032-x.

[72] Telesca L, Lovallo M. Analysis of seismic sequences by using the method of visibility graph, Europhysics Letters 2012;97(5): 50002, doi:10.1209/0295-5075/97/50002.

[73] Zou Y, Heitzig J, Donner RV, Donges JF, Farmer JD, Meucci R, Euzzor S, Marwan N, Kurths J. Power-laws in recurrence networks from dynamical systems, Europhysics Letters 2012;98(4): 48001, doi:10.1209/0295-5075/98/48001.

[74] Dall J, Christensen M. Random geometric graphs, Physical Review E 2002;66(1): 016121, doi:10.1103/PhysRevE.66.016121.

[75] Donges JF, Heitzig J, Donner RV, Kurths J. Analytical framework for recurrence network analysis of time series, Physical Review E 2012;85(4): 046105, doi:10.1103/PhysRevE.85.046105.

[76] Donner RV, Heitzig J, Donges JF, Zou Y, Marwan N, Kurths J. The geometry of chaotic dynamics – a complex network perspective, European Physical Journal B 2011;84(4): 653-672, doi:10.1140/epjb/e2011-10899-1.

[77] Zou Y, Donner RV, Donges JF, Marwan N, Kurths J. Identifying complex periodic windows in continuous-time dynamical systems using recurrence-based methods, Chaos 2010;20(4): 043130, doi:10.1063/1.3523304.

[78] Zou Y, Donner RV, Kurths J. Geometric and dynamic perspectives on phase-coherent and noncoherent chaos, Chaos 2012 ;22(1) : 013115, doi :10.1063/1.3677367.

[79] Donges JF, Donner RV, Trauth MH, Marwan N, Schellnhuber HJ, Kurths J. Nonlinear detection of paleoclimate-variability transitions possibly related to human evolution, Proceedings of the National Academy of Sciences 2011;108(51): 20422-20427, doi:10.1073/PNAS.1117052108.

[80] Hays JD, Imbrie J, Shackleton NJ. Variations in the Earth's orbit: Pacemaker of the ice ages, Science 1976;194(4270): 1121-1132, doi:10.1126/science.194.4270.1121.

[81] Buck CE, Millard AD, editors. Tools for Constructing Chronologies. London: Springer; 2004.

[82] Parnell AC, Buck CE, Doan TK. A review of statistical chronology models for high-resolution, proxy-based Holocene palaeoenvironmental reconstruction, Quaternary Science Reviews 2011;30(21-22): 2948-2960, doi:10.1016/j.quascirev.2011.07.024.

[83] Priestley MB. Spectral Analysis and Time Series. London: Academic Press; 1981.

[84] Ghil M, Allen MR, Dettinger MD, Ide K, Kondrashov D, Mann ME, Robertson AW, Saunders A, Tian Y, Varadi F, Yiou P. Advanced spectral methods for climate time series, Reviews of Geophysics 2002;40(1), 1003, doi:10.1029/2000RG000092.

[85] Babu P, Stoica P. Spectral analysis of nonuniformly sampled data – a review, Digital Signal Processing 2009;20(2): 359-378, doi:10.1016/j.dsp.2009.06.019.

[86] Rehfeld K, Marwan N, Heitzig J, Kurths J. Comparison of correlation analysis echniques for irregularly sampled time series, Nonlinear Processes in Geophysics 2011;18(3): 389-404, doi:10.5194/npg-18-389-2011.

[87] Lomb NR. Least-squares frequency analysis of unequally spaced data Astrophysics and Space Science 1976;39: 447-462, doi:10.1007/BF00648343.

[88] Scargle J. Studies in astronomical time series analysis. II. Stattistical aspects of spectral analysis of unevenly spaced data, Astrophysical Journal 1982;263: 835-853, doi:10.1086/160554.

[89] Scargle J. Studies in astronomical time series analysis. III. Fourier transforms, autocorrelation functions, and cross-correlation functions of unevenly spaced data, Astrophysical Journal 1989;343): 874-887, doi:10.1086/167757.

[90] Mudelsee M. Climate Time Series Analysis: Classical Statistical and Bootstrap Methods. Dordrecht: Springer; 2010.

[91] Priestley MB. Non-linear and non-stationary time series analysis. London: Academic Press; 1988.

[92] Daubechies, I. Ten Lectures on Wavelets. Philadelphia: SIAM; 1992.

[93] Holschneider, M. Wavelets: An Analysis Tool. Oxford: Oxford University Press; 1995.

[94] Percival DB, Walden AT. Wavelet Methods for Time Series Analysis. Cambridge: Cambridge University Press; 2000,

[95] Foster G. Wavelets for period analysis of unevenly sampled time series, Astronomical ournal 1996;112(4): 1709-1729, doi:10.1086/118137.

[96] Andronov IL. Method of running parabolae: Spectral and statistical properties of the smoothing function, Astronomy and Astrophysics Supplement Series 1997;125: 207-217, doi:10.1051/aas:1997217.

[97] Andronov IL. Wavelet analysis of time series by the least-squares method with supplementary weights, Kinematics and Physics of Celestial Bodies 1998;14(6): 374-392.

[98] Andronov IL. Wavelet analysis of the irregular spaced time series. In: Priezzhev VB, Spiridonov VP (eds.) Self-Similar Systems. Dubna: JINR; 1999. p57-70.

[99] Schumann AY. Waveletanalyse von Sedimentdaten unter Einbeziehung von Alters-Tiefen-Modellen (in German). Diploma Thesis. University of Potsdam; 2004.

[100] Witt A, Schumann AY. Holocene climate variability on millennial scales recorded in Greenland ice cores, Nonlinear Processes in Geophysics 2005;12(3): 345-352, doi:10.5194/npg-12-345-2005.

[101] Frick P, Baliunas SL, Galyagin D, Sokoloff D, Soon W. Wavelet analysis of stellar hronomspheric activity variations, Astrophysical Journal 1997;483: 426-434, doi:10.1086/304206

[102] Frick P, Grossmann A, Tchamitchian P. Wavelet analysis of signals with gaps, Journal of Mathematical Physics 1998;39(8): 4091-4107, doi:10.1063/1.532485.

[103] Otazu X, Ribo M, Peracaula M, Paredes JM, Nunez J. Detection of superimposed ignals using wavelets, Monthly Notices of the Royal Astronomical Society 2002;333(2), 365-372, doi:10.1046/j.1365-8711.2002.05396.x.

[104] Otazu X, Ribo M, Paredes JM, Peracaula M, Nunez J. Multiresolution approach for period determination on unevenly sampled data, Monthly Notices of the Royal Astronomical Society 2004;351(1), 215-219, doi:10.1111/j.1365-2966.2004.07774.x.

[105] Nicolis C, Nicolis G. Is there a climatic attractor? Nature 1984;311(5986): 529-532, doi:10.1038/311529a0.

[106] Nicolis C, Nicolis G. Reconstruction of the dynamics of the climatic system from time-eries data, Proceedings of the National Academy of Sciences 1986;83(3): 536-540, http://www.jstor.org/stable/27375.

[107] Grassberger P. Do climatic attractors exist? Nature 1986;323(6089): 609-612, doi:10.1038/323609a0.

[108] Maasch KA. Calculating climate attractor dimension from $\delta^{18}O$ records by the Grassberger-Procaccia algorithm, Climate Dynamics 1989;4(1): 45-55, doi: 10.1007/VF00207399.

[109] Fluegeman Jr RH, Snow RS. Fractal analysis of long-range paleoclimatic data: Oxygen sotope record of Pacific core V28-239, Pure and Applied Geophysics 1989;131(1-2): 307-313, doi:10.1007/BF00874493.

[110] Schulz M, Mudelsee M, Wolf-Welling TCW. Fractal analyses of Pleistocene marine oxygen isotope records. In: Kruhl JH (ed.) Fractals and Dynamic Systems in Geoscience. Berlin: Springer; 1994. p377-387.

[111] Mudelsee M, Stattegger K. Application of the Grassberger-Procaccia algorithm to the $\delta^{18}O$ record from ODP site 659: Selected methodical aspects. In: Kruhl JH (ed.) Fractals and Dynamic Systems in Geoscience. Berlin: Springer; 1994. p390-413.

[112] Mudelsee M, Stattegger K. Plio-/Pleistocene climate modeling based on oxygen isotope ime series from deep-sea sediment cores: The Grassberger-Procaccia algorithm and haotic climate systems, Mathematical Geology 1994;26(7): 799-815, doi: 10.1007/BF02083118

[113] Mudelsee M. Entwicklung neuer statistischer Analysemethoden für Zeitreihen mariner, stabiler Isotopen: die Evolution des globalen Plio-/Pleistozänen Klimas (in German). PhD thesis. University of Kiel; 1995.

[114] Gibson JF, Farmer JD, Casdagli M, Eubank S. An analytic approach to practical state space reconstruction, Physica D 1992;57(1-2): 1-30, doi:10.1016/0167-2789(92)90085-2.

Fractal Analysis of InterMagnet Observatories Data

Sid-Ali Ouadfeul and Mohamed Hamoudi

Additional information is available at the end of the chapter

1. Introduction

The fractal analysis has been widely used in geophysics (Ouadfeul, 2006; Ouadfeul, 2007; Ouadfeul and Aliouane; 2010a; Ouadfeul and Aliouane, 2010b; Ouadfeul and Aliouane; 2011; Ouadfeul et al; 2012a, 2012b).

The purpose of this chapter is to use the fractal analysis to detect and establish a schedule of geomagnetic disturbances by analyzing data from the International Real-time Magnetic Observatory Network (INTERMAGNET) observatories. This will be achieved by the use of fractal formalism revisited by the continuous wavelet transform. Several techniques have been applied for prediction of geomagnetic disturbances. We cite, for example, the technique of neural networks for prediction of magnetic storms (Iyemori et al, 1979; Kamide et al, 1998; Gleisner et al, 1996).

 In this chapter we analyze signals by the fractal formalism to predict geomagnetic storms and provide a schedule of geomagnetic disturbances. The technique of maximum of modulus of the continuous wavelet transform is used. We start the chapter by giving some definitions in geomagnetism. We then give a short description of magnetic storm and its effects. The proposed methods of analysis are then applied to data recorded by different observatories.

2. Overview of the geomagnetic field

2.1. Definitions

Earth's magnetic field or geomagnetic field is a complex electrodynamics phenomenon, variable in direction and intensity in space and time. Its characterization is useful to isolate the different geomagnetic field contributions (Merrill and Mc Elhinny, 1983). Recall that at the Earth's surface, one can distinguish two main processes. The first one is due to the inside of the Earth (internal field) and the second one is due to the outside (external field) (Le Mouël, 1969). The internal field is presented by the main field (99%) from the measured one at the earth's surface. The geodynamo is the origin of earth magnetic field. The lithospheric anomaly filed is due to rocks magnetization located above the Curie isothermal surface. The external geomagnetic field is generated by external electric currents flowing in the ionosphere and magnetosphere.

2.2. Modelling of the geomagnetic field

Since the nineteenth and the time of Gauss, the modelling of the internal magnetic field consists to develop a synthetic mathematical expression based on the contribution of multipolar magnetic sources. This modelling is based on observations (Chapman and Bartels, 1940). The mathematical model is used either to study the core field, or to serve as reference field for exploration and navigation on the Earth's surface.

The first description of the geomagnetic field is quoted to Gauss in 1838. He used the spherical harmonic expansion to characterize the Earth's magnetic field.

In a source-free medium free of sources (e.g., between the surface of the Earth and the lowest layer of the ionosphere), one can show that the field derives from a scalar potential V following B = - grad(V). It means that this potential satisfies the Laplace equation's $\Delta V = 0$. So the potential is harmonic and is expressed in spherical coordinates as a spherical harmonic expansion (Chapman and Bartels, 1940):

$$V(r,\theta,\varphi) = a\sum_{n=1}^{\infty}\sum_{m=0}^{n}\left(\frac{a}{r}\right)^{n+1}(g_n^m(t)\cos(m\varphi) + h_n^m(t)\sin(m\varphi))P_n^m(\cos(\theta)) + a\sum_{n=1}^{\infty}\sum_{m=0}^{n}\left(\frac{r}{a}\right)^{n}(q_n^m(t)\cos(m\varphi) + s_n^m(t)\sin(m\varphi))P_n^m(\cos(\theta)$$

| Internal Sources | External sources |

Where:

(r,θ,φ) are the spherical coordinates, t is time and a is the mean radius of the Earth (6371.2 km). n, m, are respectively the degree and the order of development. $g_n^m(t), h_n^m(t)$ are the Gauss's internal coefficients, and $q_n^m(t), s_n^m(t)$ are the Gauss external coefficients. $P_n^m(\cos(\theta))$ are the associated Legendre polynomials semi-normalized in the sense of Schmidt.

2.3. Elements of the geomagnetic field

The geomagnetic field is a vector characterized by a direction and intensity. At any point P of the space, the geocentric components B_r, B_θ and B_φ of the geomagnetic field are connected to the geographical components X, Y and Z.

The complete determination of the field at a point P in space requires the measurement of three independent elements may be chosen from the following seven elements (Fig. 1):

1. The East component $X = -B_\theta$.
2. The West component $Y = B_\varphi$.

3. The vertical component $Z = -B_r$.
4. Magnetic declination D may simply be defined as the azimuth of the magnetic meridian. The declination is positive when the magnetic meridian is east of geographic meridian.

The magnetic inclination I, is the angle between the direction of the field vector and its projection on the horizontal plane. It is positive when the vector field points towards the interior of the Earth (Northern Hemisphere).

5. The intensity B (or sometimes labelled T) or the module of the magnetic field is given by the following formula:

$$B = \sqrt{X^2 + Y^2 + Z^2}$$

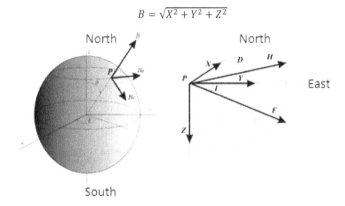

Figure 1. Elements of the geomagnetic field at a point P in the geocentric spherical coordinate system (a) and the local geographic coordinate system (b). (adapted from Blakely, 1996)

6. The horizontal component H, or any of its components horizontal, Y or X, will be particularly sensitive to external effects, such as magnetic storms.

2.4. The magnetic storms

Solar activity modulates the transient variations of the geomagnetic field. In particular, the undecennal cycle is clearly seen in the temporal distribution of sunspots as well as the magnetic activity as highlighted by the variation of the K or Dst indices. The increase in electron density due to the solar wind in different layers of the ionosphere would vary the intensity of the geomagnetic field causing many effects. The main effects (Campbell, 1997) are:

- Change the direction of the magnetic field;
- Variation of the intensity of fluctuations of the terrestrial magnetic field, mainly the horizontal component;
- Induced noise in electric cables or in the telephone;

- Disturbance of important ionospheric propagation of radio waves;
- Appearances of auroras.

There are two types of magnetic storms:

- The Storm Sudden Commencement (SSC) which is synchronous at all points of the Earth. The magnitude of the storm could reach the thousand of nT. It is more intense in the vicinity of the maxima of the solar cycles. SSCs occur few tens of hours after solar flares. It is accompanied by an intense emission of ultraviolet rays affecting the ionospheric layers D and E, in addition to showers of fast protons. The storm may last several days.
- The storm with gradual onsetwhich is characetrized by a moderate average intensity, and more localized effects. It often occurs with some regularity of the order of 27 days, corresponding to the period of intrinsic solar rotation.

Monitoring the solar activity can help to predict certain disturbances in the propagation of waves whose effects can be serious for telecommunications, as well as the impact of these storms on the distribution of electrical energy. In 1965, a massive power failure plunged the North American continent in the dark, or 30 million people out of 200 000 km2 (Campbell, 1997). In 1989, a failure of the same origin affected 6 million people in Quebec (Canada). The auroras produced by this storm were visible over Texas.

Figure 2. Interaction of the solar magnetic storm with the magnetosphere (Wékipedia).

2.5. Dst index

The geomagnetic Dst index (Menvielle, 1998) is an index that tracks the global magnetic storms. It is built by the average of the horizontal component H of the geomagnetic field measured at mid-latitude observatories. Negative values of Dst indicate a magnetic storm in progress. The minimum value of Dst indicates the maximum intensity of the magnetic storm.

3. Fractal analysis of INTERMAGNET observatories data

To analyze data from observatories, an algorithm was developed (Fig. 3). It is based on the estimation of Hölder exponents at maxima of modulus of the Continuous Wavelet Transform (CWT).

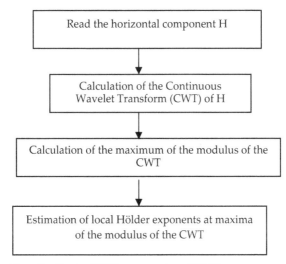

Figure 3. Flowchart for estimating local Hölder exponents on WTMM

3.1. Wingst observatory data analysis for May 2002

We analyze the Wingst magnetic observatory (Germany) data recorded during May 2002. This period saw large geomagnetic disturbances. (See tables 1 and 2). Table 3 gives all information on the location and type of magnetometer used for Wingst observatory. We analyzed the horizontal component of the magnetic field. This last is calculated from the X and Y components. Figure 4 shows this component versus the time. The modulus of the continuous wavelet transform is shown in Figure 5.

Date	Hour	Max Hour	End Hour
11	11.21	11.32	11.41
17	15.50	16.08	16.14
20	10.14	10.29	10.34
20	10.49	10.53	10.56
20	15.21	15.27	15.31
22	17.55	23/10.55	24/14.55

a

Date	Starting Hour
11	10.13
14	xx.xx
23	1.48

b

Table 1. a Major solar events that could cause a perturabation; **b** Magnetic storms recorded during the month of May 2003.

Station Code	WNG
Organization	GeoForschungsZentrum, Potsdam
Co-latitude	36,257°N
Longitude	9,073 ° E
Altitude:	50 Meter
Country	Germany

Table 2. Characteristics of the Wingst observatory

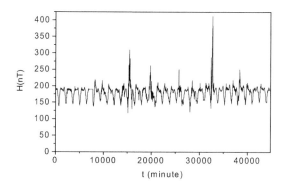

Figure 4. Horizontal component of the magnetic field recorded by the Wingst observatory during the month of May 2002.

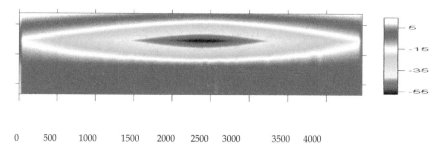

Figure 5. Wavelet Coefficients in time-log(scale) frame

The next step consists to calculate the Hölder exponents at the local maxima of the modulus of the continuous wavelet transform. Note that the calculation of Hölder exponents at maxima of the modulus of the CWT is the core of this analysis. Indeed, this is the point which can differentiate the proposed method over other methods based on the Hölder exponent estimation. This method allows us to save time by calculating the Hölder exponents only at representative times of magnetic disturbances. Figure 6 is a representation of the obtained results. To compare the obtained results with the Dst index, we calculated an average value of the Hölder exponents for each hour of the month. The obtained results are shown in Figure7.

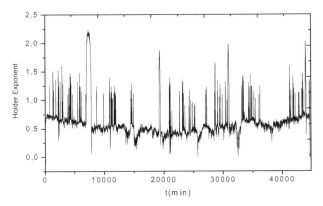

Figure 6. Estimated Hölder exponents at maxima of the modulus of the CWT.

Results and interpretation

It is clear that the horizontal component H is characterized by a Hölder exponent of very low value at the moment of the magnetic storm (Figure 7). Each event is characterized by a peak in the curve of the Dst index. Figure 8 is a detailed presentation between days 8 to 13, showing the behaviour of the Hölder exponent before, during and after the magnetic storm. We observe that hours before the storm are characterized by progressive decrease of the Hölder exponent to reach the minimum value h = 0.07 at t = 11.55day (11th day and 13.20 hours). After the storm, we observe a gradual increase of the Hölder exponent.

3.2. Baker Lake observatory data analysis

We analyzed the horizontal component recorded by the Baker Lake Observatory for May 2002. The observatory's informations are given in table 4.

Figure 9 presents the horizontal component of the magnetic field and figure 10 presents the average local Hölder exponents for each hour of the month compared with the Dst index. We note that the major magnetic storms are characterized by negative values of the Dst index and by very low values of the Hölder exponents.

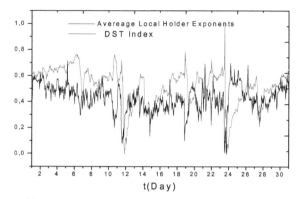

Figure 7. Average Hölder exponents compared with the Dst index .

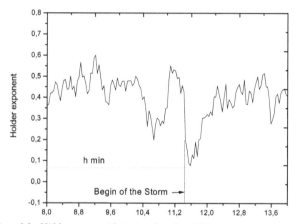

Figure 8. Variation of the Hölder exponent between days 8 and 13

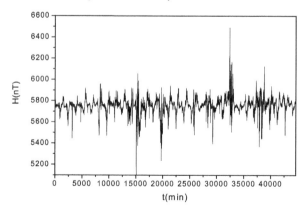

Figure 9. Horizontal component recorded by the Baker Lake Observatory during May 2002.

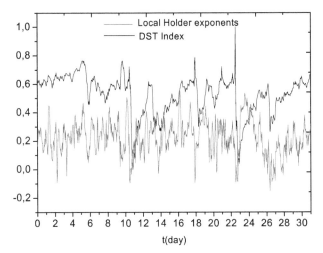

Figure 10. Local Hölder exponents mean calculated every hour and compared to the Dst index.

4. Demonstration of the multifractal character of the signal of the external geomagnetic field

The WTMM can distinguish between monofractal and multifractal (Arnéodo et al., 1988; Arnéodo and Bacry, 1995). We will use this feature to demonstrate the multifractal character of the external geomagnetic field. Data analysis of the total field recorded by the Wingst observatory during the month of May 2002 by the WTMM technique gives a spectrum of exponents and a spectrum of singularities that demonstrate the multifractal character of the external geomagnetic field signals. (See figure11). Indeed, the spectrum of exponents is not linear and the spectrum of singularities is not concentrated at one point.

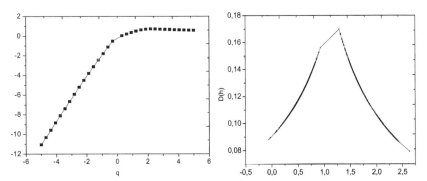

Figure 11. Multifractal analysis of the external geomagnetic field for the period of May 2002.
(a) Spectrum of exponents. (b) Spectrum of singularities.

5. The generalized fractal dimensions as an index of magnetic disturbances

In this section we will demonstrate the usefulness of generalized fractal dimension D(q) to establish a calendar of magnetic disturbances. We will apply this technique at the data of many InterMagnet observatories data. Two important periods are analyzed. One is the month of May 2002 and the second is the period of October and November 2003.5.1 Data Analysis of the period of May 2002

5.1.1. Analysis of data of Hermanus observatory

The first record to be processed is the total field recorded by the Hermanus observatory during the period of May 2002. The total field variations are shown in Figure 12.

Station Code	HER
Localisation	Hermanus
Organisation	National Research Foundation
Co-latitude	124,425°
Longitude	019,225°E
Altitude	26 m
Country	South Africa

Table 3. Characteristics of Hermanus observatory.

$$D(q) = \frac{\tau(q)}{(q-1)}$$

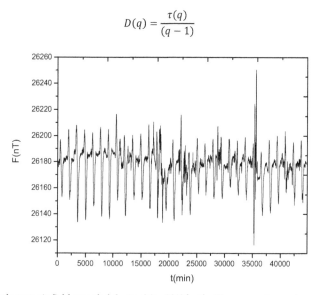

Figure 12. Total magnetic field recorded during May 2002 by the Hermanus observatory.

The first step is to apply the WTMM technique on data of each hour of the month, that is to say, every 60 samples. The objective is to estimate the spectrum of exponents τ (q). Then we compute the generalized fractal dimensions for the following values of q: 0, and 2. The following formula is used (Ouadfeul et al, 2012):

$$D(q) = \frac{\tau(q)}{(q-1)}$$

Note that for D (1) we use the limit of D (q) when q tends towards 1.

Figure 13 presents of the flow chart of the proposed technique.

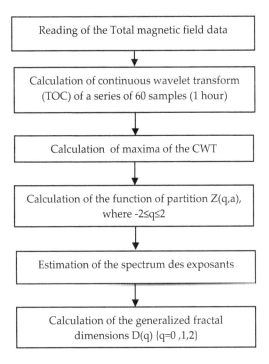

Figure 13. Flowchart of the total magnetic field analysis using the generalized fractal dimensions.

Application of the WTMM method at the first 60 samples of the total field is shown in figure 14.

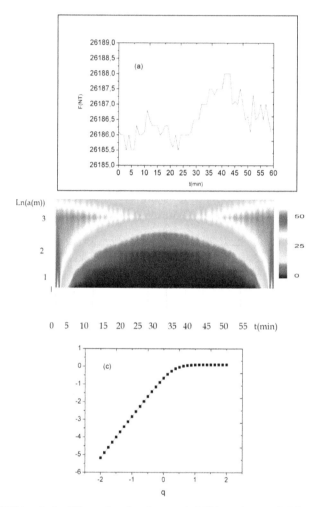

Figure 14. WTMM analysis of 60 samples of total magnetic field intensity recorded during May 2002 at Hermanus observatory.
(a) Recorded signal. (b) Wavelet coefficients.(c) Spectrum of exponents.

The same operations are applied to each hour of May 2002. Fractal dimensions are then calculated. The obtained results are shown in figure 15

Results and interpretation

Obtained results show the non sensitivity of the fractal dimension D_0 to the magnetic disturbances. However, for the generalized fractal dimensions D_1 and D_2, the main magnetic disturbances are characterized by peaks (see Tables 1a and 1b and figure 14).

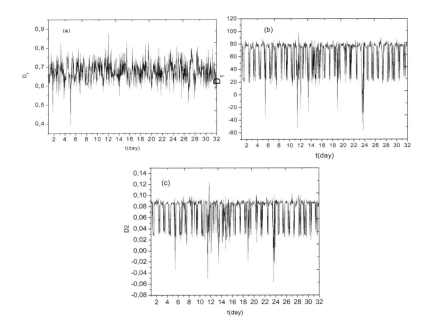

Figure 15. Fractals dimensions estimated at every hour of May 2002 (Hermanus Observatory)
(a) D_0, (b) D_1, (c) D_2

The obtained results show that peaks are observed in the generalized fractal dimensions D_1 and D_2 at the times of occurrences of magnetic storms. These dates correspond to 11, 14 and 23 of May 2002. One can notice that the dimension D_0 is not clearly sensitive to the magnetic storm.

5.1.2. Baker Lake observatory data analysis

We analyzed by the same way the signal of magnetic field intensity recorded by the Baker Lake Observatory during May 2002. Figure 16 shows the fluctuations of this field with time.

The generalized fractal dimensions are calculated at each hour of the month. The obtained results are shown in Figure 17. We note that peaks are observed in the plot of D_1 and D_2 at the time of the magnetic storms (Tables 1a and 1b).

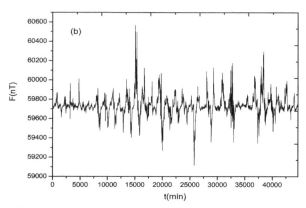

Figure 16. Total field recorded during May 2002 by the Baker Lake observatory

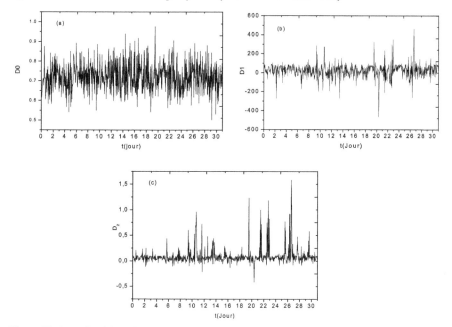

Figure 17. Generalized fractal dimensions calculated for each hour of the month of May 2002, of the Baker Lake observatory (a) D_0 (b) D_1 (c) D_2

5.2. Analysis of geomagnetic data for months of October and November 2003

5.2.1. Data analysis from the Kakioka observatory

We analyzed the total field recorded by the Japanese Kakioka observatory, the details of this observatory are summarized in table5. Figure 18 shows the total magnetic field fluctuations during the months of October and November 2003. The generalized fractal dimensions are shown in figure 19.

Station Indicative	KAK
Localisation	Kakioka, Japan
Organisation:	Japan Meteorological Agency
Co-latitude:	53,768°
Longitude:	140,186°E

Table 4. Characteristics of the Kakioka observatory

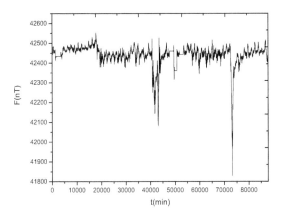

Figure 18. Total field recorded by the Kakioka observatory during October and November 2003

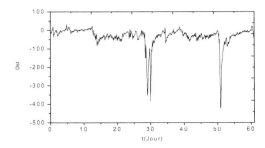

Figure 19. Dst Index calculated during the period of October and November 2003

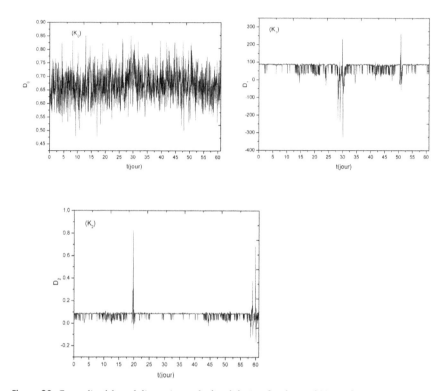

Figure 20. Generalized fractal dimensions calculated during October and November 2003. $K_1 : D_0$; K_1 : D_1 ; $K_2 : D_2$

Results analysis shows that in moments of the magnetic storm we observe significant spikes in the graphs of generalized fractal dimensions D_1 and D_2 (see figure 19 and tables 5 and 6) . The generalized fractal dimension D_0 is not sensitive to geomagnetic disturbances.

Date	Begin Hour
14	18.23
19	xx.xx
21	16.53
24	xx.xx
29	06.09
30	21.16

Table 5. Magnetic storms recorded during October 2003

Date	Begin Hour
04	06.24
06	19.33
09-18	xx.xx
20	08.02
23	16.34

Table 6. Magnetic storms recorded during November 2003

5.2.2. Data analysis of the Hermanus observatory

We analyzed by the same way, data of Hermanus observatory. Figure 21 shows the total field recorded during the months of October and November 2003. The generalized fractal dimensions are calculated using the WTMM method. Figure 22 shows the fluctuations of fractal dimensions estimated by multifractal analysis method. The same phenomena are observed in the plots of generalized fractal dimensions.

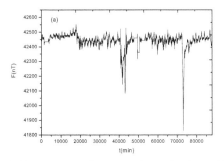

Figure 21. Total field recorded by Hermanus observatory during October and November 2003

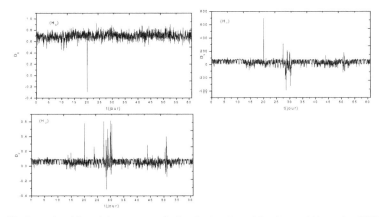

Figure 22. Generalized fractal dimensions calculated using data of October and November 2003 (Hermanus Observatory).

5.2.3. Analysis of data from the Alibag observatory

We also analyzed data recorded during the months of October and November 2003 by the Indian observatory Alibag,. Table 7 presents informations about this INTERMAGNET observatory. The recorded total field is shown in figure 23. The fractal dimensions are shown in figure 24.

Station (ID)	ABG
Localisation	Alibag
Organisation	Indian Institute of Geomagnetism (India)
Co-latitude	71.380°
Longitude	72.870°E
Country	India

Table 7. Characteristics of the Alibag Observatory

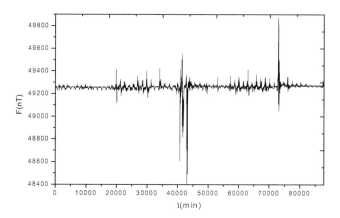

Figure 23. Total field recorded at Alibag observatory during October and November 2003

5.2.4. Analysis of data from the Wingst observatory

We analyzed the total field recorded by the Wingst observatory during October and November 2003. Figure 25 is presents the fluctuations of this field. The generalized fractal dimensions are shown in Figure 26. One can easily observe spikes in the generalized fractal dimensions at the times of magnetic storms. The analysis shows that the fractal dimensions D_1 and D_2 are very sensitive to geomagnetic disturbances. However the capacity dimension D_0 is not sensitive to magnetic disturbances. (See figure.26 and tables 5 and 6).

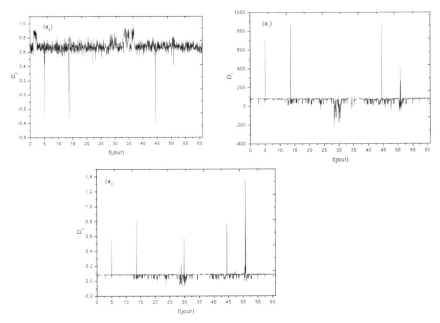

Figure 24. Generalized fractal dimensions calculated during October and November 2003 (Alibag Observatory).

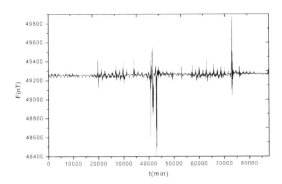

Figure 25. Total Magnetic field recorded by Wingst Observatory during October and November 2003.

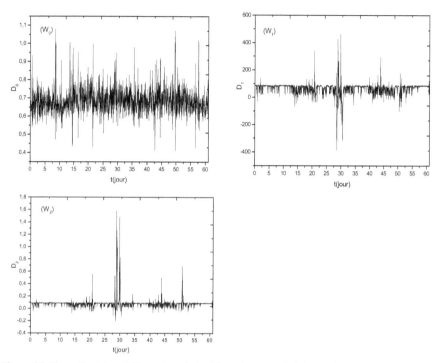

Figure 26. Generalized fractal dimensions derived from data recorded during the months of October and November 2003 (Wingst Observatory). $W_1 : D_0 ; W_1 : D_1 ; W_2 : D_2$

6. Results, discussion and conclusion

During a magnetic storm, the Hölder exponent of the horizontal component H has a very low value. Before the storm we observe gradual drop of the Hölder exponent. Several numerical experiments realized using vertical component Z recorded by magnetic observatories show that, as expected, this component is not sensitive to the magnetic disturbances. We show however that the generalized fractal dimensions D_1 and D_2 of the total magnetic field B can be confidently used as an index to describe the external magnetic activities. WTMM analysis of the horizontal component recorded during May 2002 show that this last exhibits a multifractal behaviour. This is in accordance with the analysis shown in (Ahn et al, 2007; Bolzan et al., 2009; Bolzan and Rosa, 2012). We note that these fractal dimensions show more details of solar activity compared to the Dst index. The WTMM method shows clearly the multifractal character of the signal of the geomagnetic field.

Author details

Sid-Ali Ouadfeul
Geosciences and Mines, Algerian Petroleum Institute, IAP, Algeria
Geophysics Department, FSTGAT, USTHB, Algeria

Mohamed Hamoudi
Geophysics Department, FSTGAT, USTHB, Algeria

7. References

[1] Blakely, R. J., Potential Theory in Gravity & Magnetic Applications, Cambridge University Press, 1996.

[2] Iyemori. T., Maeda. H. and Kamei. T. (1979). Impulse response of geomagnetic indices to interplanetary magnetic fields, J. Geomag. Geoelectr., 31, 1-9.

[3] Kamide. Y, Yokoyama. N, Gonzalez. W, Tsurutani. B. T, Daglis. I. A, Brekke. A., and Masuda. S.(1998)., Two step development of geomagnetic storms, J. Geophys. Res., 103, 6917– 6921.

[4] Gleisner. H and Lundstedt. H, (1997). Response of the auroral electrojets to the solar wind modeled with neural networks, J. Geophys. Res.,102, 14269–14278.

[5] Merrill, R.T., and Mc Elhinny, M.W., The Erath's Magnetic Field, Its History, Origin and Planetary Perspective, Academic Press, London (1983).

[6] Menvielle. M, (1998). Derivation and dissemination of geomagnetic indices, Revista Geofisica, 48, 51-66, 1998.

[7] Campbell, K. W. (1997), Empirical near-source attenuation relationships for horizontal and vertical components of peak ground acceleration, peak ground velocity, and pseudo-absolute acceleration response spectra, Seismological Research Letters, 68(1), 154 -179.

[8] Ouadfeul, S., (2006). Automatic lithofacies segmentation using the wavelet transform modulus maxima lines (WTMM) combined with the detrended fluctuation analysis (DFA), DFA, 17 [the] International geophysical congress and exhibition of turkey , Expanded abstract .

[9] Ouadfeul, S., (2007). Very fines layers delimitation using the wavelet transform modulus maxima lines WTMM combined with the DWT, SEG SRW.

[10] Ouadfeul, S., Aliouane, L, (2010a). 3D Seismic AVO Data Established by the Wavelet Transform Modulus Maxima Lines to Characterize Reservoirs Heterogeneities, 72nd EAGE Conference & Exhibition.

[11] Ouadfeul, S., Aliouane, L., (2010b). Wavelet Transform Modulus Maxima Lines Analysis of Seismic Data for Delineating Reservoir Fluids, Geo2010, Expanded Abstract.

[12] Ouadfeul, S., Aliouane, L., (2011). Multifractal Analysis Revisited by the Continuous Wavelet Transform Applied in Lithofacies Segmentation from Well-Logs Data, International Journal of Applied Physics and Mathematics vol. 1, no. 1, pp. 10-18.

[13] Ouadfeul, S., Hamoudi, M., Aliouane, L., (2012a). A wavelet-based multifractal analysis of seismic data for facies identification. Application on the pilot KTB borehole, Arabian Journal for Geosciences.

[14] Ouadfeul, S., Aliouane, L., Hamoudi, M., Boudella, A., and Eladj, S., (2012b). 1D Wavelet Transform and Geosciences, Wavelet Transforms and Their Recent Applications in Biology and Geoscience, Dumitru Baleanu (Ed.), ISBN: 978-953-51-0212-0, InTech.

[15] Blakely, R. J., (1996) Potential Theory in Gravity & Magnetic Applications, Cambridge University Press.

[16] Le Mouël. J.-L, 1969, Sur la distribution des éléments magnétiques en France. Thèse de la Faculté des Sciences de l'Univ. de Paris.

[17] Anh, V., Yu, Z.-G., and Wanliss, J. A., (2007), Analysis of global geomagnetic variability, Nonlin. Processes Geophys., 14, 701-708, doi:10.5194/npg-14-701-2007.

[18] Bolzan, M. J. A., Rosa, R. R., and Sahai, Y., 2009, Multifractal analysis of low-latitude geomagnetic fluctuations, Ann. Geophys., 27, 569-576, doi:10.5194/angeo-27-569.

[19] Bolzan, M. J. A. and Rosa, R. R., (2012), Multifractal analysis of interplanetary magnetic field obtained during CME events, Ann. Geophys., 30, 1107-1112, doi:10.5194/angeo-30-1107-2012.

Chapter 3

Analysis of Fractal Dimension of the Wind Speed and Its Relationships with Turbulent and Stability Parameters

Manuel Tijera, Gregorio Maqueda, Carlos Yagüe and José L. Cano

Additional information is available at the end of the chapter

1. Introduction

The atmospheric fluxes in the boundary layer at large Reynolds numbers are assumed to be a superposition of periodic perturbations and non-periodic behavior that can obey an irregular state and variable motion that is referred as turbulence. Turbulence can be observed in time series of meteorological variables (wind velocity, for example). The analysis of these series presents a self-similarity structure, [1]. So, the wind velocity can be seen as a fractal magnitude. Fractal Dimension (FD) is an artifice that shows in some way the complexity of the time series and the variability degree of the physical magnitude including. Fractal dimensions will is correlated with the characteristic parameters of the turbulence. The non-integer values of the FD are assigned to time series which exhibit a self–similarity geometry and which show that structure on all length scales. In general, turbulent flows allowed us to recognize the coexistence of structure and randomness, they are a set of solution that are not unique or depend sensitively on initial conditions [2]. The structures of these flows are related with fractal geometry.

A structure or time series is said to be self–similar if it can be broken down into arbitrarily small pieces, each of which is a small replica of the entire structure. There is a way measure this degree of complexity by means of FD. The concept of dimension is no easy to understand probably to determine what dimension means and which properties have been one of the big challenges in Mathematics. In addition, mathematicians have come up with some tens of different notions of dimension: topological dimension, Hausdorf dimension, fractal dimension, box-counting dimension, capacity dimension, information dimension, Euclidean dimension, and more [3, 4]. They are all related. We focus in the fractal dimension of these series by means of dividing its structure onto a grid with size L, and counting the number of grid boxes which contain some of the structure of series, N. This number will be depending on the size L. To obtain the FD it is needed to represent in a diagram the logarithm of N against the logarithm of the reciprocal of L (log (N(L)) versus log (1/L)). Then, the slope of the better linear fit or the linear regression between them corresponds to the searched fractal dimension or box-counting dimension. This dimension proposes a systematic measurement of degree of similarity or complexity of wind series. Various works have approached the problem of calculating the dimension associated with time series [5, 6].

In this Chapter it is going to study the Fractal Dimension (FD) of u and w fluctuations of time series of velocity recorded close to the ground, it is to say, into the low Planetary Boundary Layer, more specifically into the Surface Layer, closer to the ground and where the dynamic influence is predominant. The main aim is to seek physical quantities included in turbulent flows that correlate better with FD. Nevertheless, thermal, dynamical and combined parameters will be explored to complete the Chapter. The central idea is that the kind of stability controls the turbulent fluxes and it is showing in the values of the FD also. Differences of potential temperatures between two levels make the Surface Layer or the Low PBL unstable or stable depending on its sign. Shears in the wind produce dynamical instability making it easier the mix of properties, physical variables and mass. In neutral conditions mixture is completed and turbulence is well developed. The values of Fractal Dimension must be in accordance with the turbulent level.

It has been calculated the fractal dimension, d (Kolmogorov capacity or box-counting dimension) of the time series of the velocity component fluctuations u' and w' ($u' = u - U$, $w' = w - W$) with U the horizontal mean velocity, both in the physical space (velocity-time) [2]. It has been studied the time evolution of the fractal dimension of the u' and w' components (horizontal and vertical) of wind velocity series during several days and three levels above the ground (5.8 m, 13.5 m, 32 m).

This study is focused on the simplest boundary layer kind, over a flat surface. So, we could assume that the flow to be horizontally homogeneous. Its statistical properties are independent of horizontal position; they vary only with height and time [1, 7, 8]. The experimental data have been taken in a flat terrain with short uniform vegetation. It allows us to take on this approximation of horizontal homogeneity and on this context we focus this study on variation of fractal dimension of the horizontal and vertical components of the velocity of flow turbulent in the diurnal cycle versus to a variety of turbulent parameters: difference of potential temperatures in the layers 50-0.22 m and 32 – 5.8 m, Turbulent Kinetic Energy, friction velocity and Bulk Richardson number.

It has been observed that there is a possible correlation between the fractal dimension and different turbulent parameters, both from dynamical and thermal origin: turbulent kinetic energy, friction velocity, difference of temperature between the extreme of the layer studied close to the surface ($\Delta T_{50-0.22m}$). Finally, it has been analysed the behaviour of fractal dimension versus stability evaluated from the Richardson number.

The knowledge of turbulence and its relationships with fractal dimension and some turbulent parameters within the Planetary Boundary Layer (PBL) can help us understand how the atmosphere works.

2. Data

The data analysed was recorded in the experimental campaign SABLES-98 at the Research Centre for the Lower Atmosphere (CIBA in the Spanish acronyms), located in Valladolid province (Spain). SABLES-98 was an extensive campaign of measures in the PBL with a large number of participant teams and took place from 10th to 27th September 1998 [9]. The experimental site is located around 30km NW from Valladolid city, in the northern Iberian Plateau, on a region known as Montes Torozos, which forms a high plain of nearly 200 km^2 elevated above the plateau. The surrounding terrain is quite flat and homogeneous, consisting mainly on different crop fields and some scattered bushes. The Duero river flows along the SE border of the high plain, and two small river valleys, which may act as drainage channels in stable conditions, extends from the lower SW region of the plateau.

The main instrumentation available at CIBA is installed on the 100m meteorological tower, and includes sonic and cup anemometers, platinum resistance thermometers, wind vanes, humidity sensors, etc. Other instruments are spread in minor masts and ground based in order observes the main meteorological variables of the PBL. In this work we study five minute series from the sonic anemometers (20 Hz) installed at 5.8 (~ 6), 13.5 (~13) and 32m were used to evaluate the fractal dimension and turbulent parameters. These series have been obtained once we have carried out the necessary transformation to get the mean wind series in short intervals, namely 5 minutes, to ensure the consistent properties of turbulence [10].

We focus this study in a period of eight consecutive days (from 14 to 21 September 1998) in which have been analyzed every records of the velocity fluctuations. The synoptic conditions were controlled by a high pressure system which produces thermal convection during the daily hours and from moderate to strong stable stratification during the nights. The evolution of wind speed, Bulk Richardson number (Rib), Turbulent Kinetic Energy (TKE), friction velocity ($u \cdot$), potential temperature difference between 32m and 5.8m, and the temperature difference near surface but in an deeper layer (50-0.22 m, named invT$_{50-0.22}$, in reference to thermal inversion), with fractal dimension of velocity u and w component fluctuations at three levels above the ground (5.8 m, 13.5 m, 32 m) are analysed.

3. Methodology

The turbulent flows show very high irregularity for wind velocity time series. They present a self–similarity structure, it is to say, that for different scales the structure of the variables remains similar, as it is shown in Fig. 1. This property of the turbulent flows is related with the Fractal Dimension, since the irregularity is a common characteristic. Their non-integer values can help us analyze how the irregularity of the sign is, as well as of its geometry. As the bigger the values of FD, the more irregularity and random is the flow.

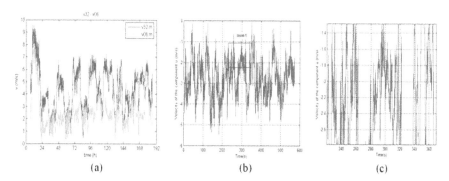

(a)   (b)   (c)

Figure 1. a) Variation of wind speed versus time from 14 to 21 September 1998, at 5.8 m and 32 m. b) If we zoom red window, in figure c), it is observed that the structure of this flow is similar over all scales of magnification.

In this section we describe the methodology applied to calculate the Fractal Dimension d (Kolmogorov capacity or box-counting dimension) [11].The more precise definition of the fractal dimension are in Hausdorff's work, become later known as Hausdorff dimension [12]. This dimension is not practical in the sense that it is very difficult to compute even in elementary examples and nearly impossible to estimate in practical applications. The box-counting dimension simplifies this problem, being an approximation of the Hausdorff dimension and is calculated approximately by

$$d = \lim_{L \to 0} \frac{\log N(L)}{\log\left(\frac{1}{L}\right)} \tag{1}$$

$N(L)$ is the number of the boxes of side L necessary to cover the different points that have been registered in the physical space (velocity-time) [13]. As $L \to 0$ then $N(L)$ increases, N meets the following relation:

$$N(L) \cong kL^{-d} \tag{2}$$

$$\log N(L) = \log k - d \log L \tag{3}$$

By means of least – square fitting of representation of log $N(L)$ versus log L, it has been obtained of the straight line regression given by equation (3), as is shown in Fig 2. The fractal dimension d will come given by the slope of this equation.

4. Stability of stratification and turbulence

The origin of turbulence cannot be easily determined, but it is know that both the dynamic and the thermal effects contribute strongly to turbulence by producing a breakdown of streamlined flow in a previously nonturbulent movement. The dynamical effects are represented by wind shear production and the thermal ones make differences of density in the fluid giving rise to hydrostatic phenomena and buoyance.

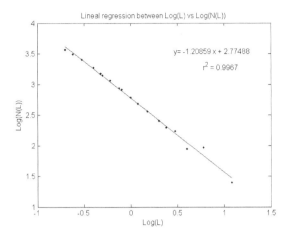

Figure 2. Example of linear regression between number of boxes not empty and length side of the box. The slope (d) is the fractal dimension, d =1.21 ± 0.02 for the example case.

The Richardson number is a parameter that includes both dynamic and thermal effects to measure the degree of stratification stability in the low Atmosphere. The static stability parameter, $s = (g / \theta)(\partial\theta / (\partial z))$, only takes in mind the buoyancy, i.e., the thermal effects. Nevertheless, a ratio between s and squared wind shear $(\partial V/\partial z)^{-2}$, gives a nondimensional product more appropriate for to calculate the stability. This ratio is known as gradient Richardson number. For the most practical cases a needed numerical approximation will be introduced below.

In the low Planetary Boundary Layer the atmosphere responds to changes in stratification stability brought about by the heating and cooling of the ground. We search the behaviour or relation between the Fractal Dimension and a parameter to establish a proper measure of stability in the surface layer. One of the most widely used indicator of stability close to the ground in atmospheric studies is the Bulk Richardson number Ri_B, a nondimensional parameter representing the ration of the rate of production or destruction of turbulence by buoyancy to that by wind shear strain is caused by mechanical forces in the atmosphere:

$$Ri_B = \frac{g}{\theta}\frac{\overline{\Delta\theta}\Delta z}{(\overline{\Delta u})^2} \tag{4}$$

where g is the gravity acceleration and $\bar{\theta}$ the average potential temperature a the reference level, the term $\frac{g}{\theta}$ is referred to as the buoyancy parameter. Ri_B is positive for stable stratification, negative for unstable stratification and approximate to zero for neutral stratification, [10, 14, 15]. The way to calculate this number is following:

1. Calculate the mean potential temperatures at height $z = 32$ m, and the close surface $z = 5.8$ m, namely $\overline{\theta_{32}}$ and $\overline{\theta_{5.8}}$ respectively. Being $\Delta\overline{\theta} = \overline{\theta_{32}} - \overline{\theta_{5.8}}$.

2. Obtain $\overline{u_z}$ the module mean wind velocity at height $z = 32$ m and $z = 5.8$ m, denoted $\overline{u_{32}}$ and $\overline{u_{5.8}}$ respectively, where $\Delta\overline{u} = \overline{u_{32}} - \overline{u_{5.8}}$.

Once obtained the values of $\Delta\overline{\theta}$, $\Delta\overline{u}$ and Δz by means of the Eq. 4 calculate the Bulk Richardson number in a layer 32m to 5.8m. The mean properties of the flow in this layer as the wind speed and temperature experience their sharpest gradients. In order to know the influence of the kind of stratification over de fractal structure of the flow in the PBL, we going to analyse the behaviour of FD and its possible changes versus the parameter Ri$_B$, as the better parameter of stability obtained from the available data.

5. Results

In this section we present the variation of the Fractal Dimension of the u', horizontal, and w', vertical, components of the velocity fluctuations along the time at the three heights of the study: 5.8 m, 13.5 m and 32 m. We observe that these variations in the three heights are similar. The daily cycle during the period of study is clearly shown in Fig. 3. No significant different values are observed in levels, but a light increasing seems outstanding in diurnal time at the lower level (red line corresponds to 5.8m above the ground). Two components, u' and w', present no differences in the time evolution.

As is shown in the Fig. 3, it is observed that the variation interval values of the fractal dimension range between 1.30 and nearly to 1.00. During the diurnal hours the fractal dimension is bigger than at night. A subtle question that concerns us is to what owes this. A possible explanation is that fractal dimension is related with atmospheric stability and, by the same reason, with the turbulence. It is well known that intensity of the turbulence grows as solar radiation increase, producing instability close the ground. In other hand, it is observed that Fractal Dimension is lightly inferior for stable stratification. We shall come back to this matter forward. In the nights a strong stability atmospheric usually exists, so the fractal dimension is usually smaller than during the diurnal hours.

5.1. Potential temperature and fractal dimension

This section concerns with the exam of the relation between the potential temperature differences and the Fractal Dimension. Potential temperature is a very useful variable in the Planetary Boundary Layer that can be replace the observed temperature in the vertical thermal structure, since an air parcel rises or goes down adiabatically at potential temperature constant. A vertical profile of potential temperature uniform represents a neutral stratification or it is called as adiabatic atmosphere.

Next, we show the variation of potential temperature at heights $z = 32$ m and $z = 5.8$ m along the time in Fig. 4. The features shown in this figure need some comment. The potential temperature at height $z = 5.8$ m is bigger than to 32 m in the nights. It is a characteristic of the nocturnal cooling that produces inversion. During the noon the potential temperatures decrease slightly with height. It is corresponding to instable conditions in the surface layer and we have a possible mixture of both mechanical and convective turbulence [16]. However, the evolution of potential temperature from minimum value until maximum is identical for the two levels in every day studied. It is the consequence of neutral situation with efficient mix during the morning.

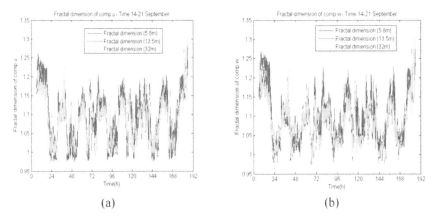

(a)   (b)

Figure 3. The variation of the Fractal Dimension (FD) *versus* time present an analogous behavior for the u' and w' component fluctuations ((a) and (b), respectively) at the 3 levels, showing the influence of the diurnal cycle.

In the Figure 4, it is clear to observe the diurnal cycle, observing a strong thermal inversion during the nights that is starting after noon and increasing uniformly to reach the maximum value. The main differences between potential temperatures are observed under inversion condition as a result of the separation among layers because the strong stratification.

In the Fig. 5, it is shown the fractal dimension at three heights and the difference of potential temperature between 50 and 0.22 m. In this study we will name it as thermal inversion because positive values correspond to actually inversion, in terms of potential temperature ($invT_{50-0.22}$). The layer 0.22 to 50 m is more extensive than the stratum defined by the levels of the sonic anemometers covering a deep part of the PBL and likely the whole Surface Layer. Besides the instruments to measure the invT are others different of the sonic anemometers which can to give a temperature by the speed of the sound measured also in they. This has an evident benefit in the results.

It is observed that fractal dimension correlates in opposed way with the difference of potential temperatures in the layer between 50 and 0.22 m. The more stable conditions are coincident with the bigger values of fractal dimension. In strong thermal inversions the fractal dimensions are lower. Although this good correlation is observed in the figure, we can justify it by mean of a least–squares fitting between the data in the different temporal intervals corresponding to the variables analyzed.

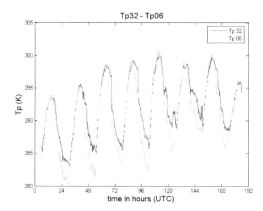

Figure 4. Time series of potential temperature at 32 and 5.8 m along the complete period of study.

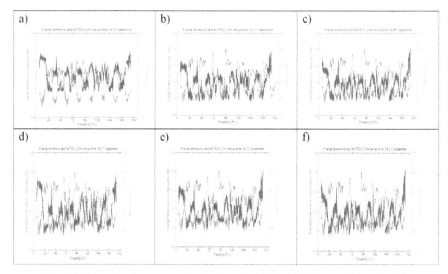

Figure 5. Fractal dimension of component fluctuations u' and w' and the thermal inversion in layer between 0.22 and 50 m versus time at three heights: z=5.8 m, z=13 m and z=32 m

As it is observed in the Fig. 6, we show the linear regression in two temporal intervals corresponding to the first 24 hours -from 06 UTC of first day to 06 UTC of the second day- and at night interval from 18 UTC to 06 UTC. The values obtained of the correlation coefficient are very good, in the first case R = -0.925, in the second one R = -0.707. It is observed a better correlation during 24 hours than the night hours because the variation of the invT and FD are wider along all day. The fractal dimension is bigger during the day time than during the night, outstanding a strong inversion at night hours. We could achieve similar results in others temporal intervals of the two variables analysed in the period study. These results are obtained for fluctuations of w' component of velocity at height z = 5.8 m. Similar results are obtained for other two heights z =13 m and z = 32 m, also for fluctuations of u' component, along of the mean wind direction at three heights studied.

In order to extent the study to all days of the experimental campaign it has been analysed the relation between the Fractal Dimension and the difference of potential temperatures using averages of FD in defined intervals of invT. Figure 7 presents the cloud of the points for all values measured in 5 minutes interval for FD and InvT with the average and the error bars based on standard deviation. Both u' as w' have similar behaviour respect the variation of invT. In these cases FD is calculate for the lowest level, i. e. 5.8 m (6m labelled in the figure). An analogous shape of the cloud of points and identical results can be obtained if the potential temperature differences when the stratum 32-5.8m has been used. It is not shown in this Chapter because repetitive.

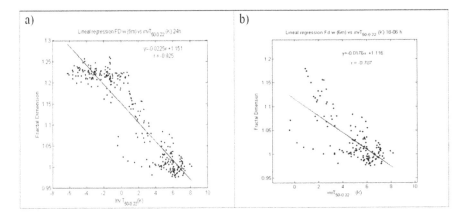

Figure 6. As it is observed the fit is very good in the temporal intervals with r satisfactory, it is showed the linear regression between the fractal dimension and la difference of the temperature between two levels in the periods of 24 h and 18-06h ((a) and (b) respectively).

It is observed that in two components a similar behaviour. Fractal Dimension have bigger values in unstable and near neutral conditions (negative and close to zero invT), reaching fractal dimension average near 1.15 value for the interval from -1 to 1 K of temperature differences. For strong inversion (> 1 K) the fractal dimension becomes smaller, around 1.05. The presented figures and values can be extent to FD of u' and w' fluctuations at others two heights z = 13 m and z = 32 m with similar results.

Finally, the good relation between FD and invT has been valued by mean a linear regression shown in Figure 8, where it has been used the average value of the scatter plot in Figure 7. It is notable the negative slope (- 0.029 –in the corresponding units -) of the regression and the high correlation coefficient, R = -0.93.

a)　　　　　　　　　　　　　　　　　　b)

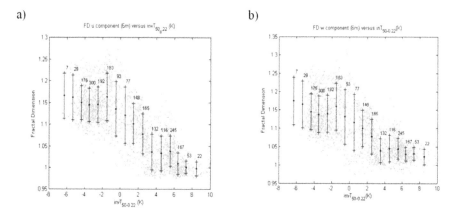

Figure 7. Scatter diagram of FD versus InvT50-0.22. It is observed a similar behavior of the fractal dimension with the inversion of temperature, which is evaluated by difference of potential temperature at 50 – 0.22m layer. Average points and error bars are shown for a better understanding.

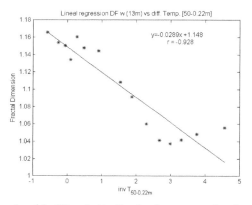

Figure 8. Linear regression of the FD against invT, using the average values from the cloud of points in Fig 7. Negative slope and the high correlation coefficient show a good correlation.

5.2. Turbulent Kinetic Energy and fractal dimension

The Turbulent Kinetic Energy (TKE) correspond to the quantity of energy associated to the movement of the turbulent flow and it is evaluated from the variances of the components of velocity: $\overline{u'^2}$, $\overline{v'^2}$ and $\overline{w'^2}$. TKE is given by Eq. 5, in terms of energy per unit mass [15].

$$TKE = \frac{1}{2}\left(\overline{u'^2} + \overline{v'^2} + \overline{w'^2}\right) \tag{5}$$

In this Section we will examine the relationship between TKE and Fractal Dimension of the fluctuations u' and w'. Before incoming to study with certain details of this relationship, it would be interesting to relate the TKE to different kind of atmospheric stability. The budgets of TKE in the instable, stable and neutral conditions near the surface are next summarized.

In the stable case, the production of the TKE by shear is not sufficient to balance the dissipation energy at all levels and production by buoyancy is not happening. So, turbulence decay and TKE decrease along the time. In neutral conditions near the surface in agreement with the observations suggest that exists a near balance between shear production and viscous dissipation [14, 17]. In this case, a strong shear wind is normally present in the low atmosphere. In the unstable layer, convective case, vertical gradients of w are bigger than the vertical gradients of u and v, the TKE production is mainly due to buoyancy and transport from other levels. Turbulence is supported by thermal effects in the case of instability [17].

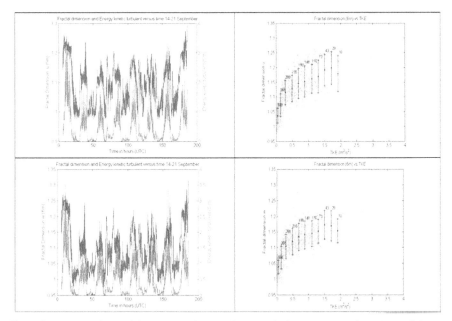

Figure 9. Variation of Fractal dimension and Energy kinetic turbulent versus time for u (upper-left panel) and w (lower-left) component of wind velocity at height $z = 5.8$ m. Scattered diagram of the fractal dimension and TKE with averages values and error bars (right panels).

As results it has found that an strong correlation between the dynamic magnitude, the Turbulent Kinetic Energy, and the Fractal Dimension exists. Figs. 9 and 10, in the left panels, show the evolution of the FD and TKE along the time for all days studied at two levels (5.8m Fig 9 and 13m in Fig 10). Diurnal cycle is also observed like in the study of invT in previous sections. This first result is in agreement with the normal variation of the turbulence between the day time and night time. Values of TKE are similar at two levels analyzed which may be interpreted as turbulence is development into the stratum in a uniform way. Right panels, in the Figs. 9 and 10, is showing the scatter plot of FD and TKE values for all data used. Averaged of FD for arbitrary intervals of TKE together its error bars are also represented. It is observed a continuous increasing of the Fractal Dimension when the TKE grows, both for the u' and w' fluctuations. The maxima values of the kinetic energy are 2 $m^2 s^{-2}$, in average, at heights z =5.8 m, z =13 m and z = 32m (do not shown) corresponds to maxima values of mean FD. Differences may be observed for FD among the figures. For u' components at 5.8 m present the highest values of FD (around 1.20 ± 0.05) while in the remainder cases w' in two levels and u' at 13m the maximum of FD are nearly to 1.15. The relation studied does not seem linear, but FD increases very fast for small values of TKE and it has a light increasing of FD beyond of 1 m^2/s^2.

In order to investigate the correlation between FD and TKE, we were tried a linear regression for two different time interval, at first, from 06 UTC to 24 UTC of the initial day, including the period of day time and early night. In this record the variation of FD is maximum. Figure 11a presents the scatter plot and the linear regression obtaining an R^2 coefficient of 0.698 that can be to consider satisfactory in spite of the shape of the cloud of points. The second period valued corresponds to the third day of the experiments from 48 UTC to 60 UTC, i.e., from midnight to noon. The scatter plot (Fig. 11b) presents similarity with the previous one, a cumulated group closes to the zero of TKE and there is spread points of higher values of FD for bigger TKE. The determination coefficient $R^2 = 0.689$ seems similar to the other period. Analogous results are obtained for u' component.

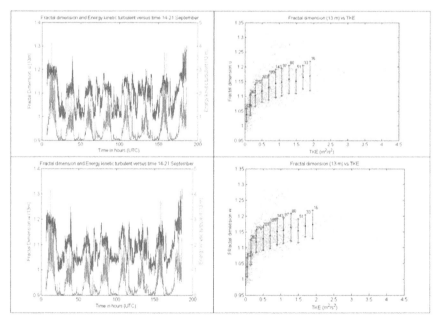

Figure 10. Variation of Fractal dimension and Energy kinetic turbulent versus time for u and w component of wind velocity at height z= 13 m. Diagram of fractal dimension and TKE with error bars

5.3. Friction velocity and fractal dimension

In the atmosphere the relevant turbulent velocity scale is the surface friction velocity u·
which includes the vertical momentum flux related by Eq. 6.

$$u_* = \left[\left(\overline{-u'w'} \right)^2 + \left(\overline{-v'w'} \right)^2 \right]^{1/4}$$

(6)

Friction velocity is also a turbulent parameter that measures the shearing stress into the
surface layer, $\tau = \tau_0 = \rho u_*^2$. It is considered constant in whole Surface Layer. So, it is an
important dynamic property of the vertical structure of the lower atmosphere. Friction
velocity can be obtained from vertical profile of average horizontal component of the wind
but also by mean of Eq. 6, based in turbulent fluxes.

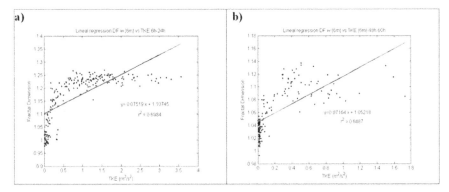

Figure 11. Linear regression between fractal dimension w component and TKE at height z=5.8 m. It is
shown the correlation coefficient in the two temporal intervals analyzed.

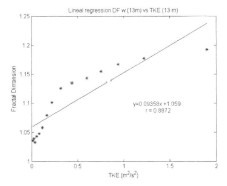

Figure 12. Average points according defined interval of TKE of Fractal Dimension of vertical
component fluctuations at 13 m level. Coefficient of correlation, $R = 0.8872$, is shown together the right
of regression.

Under neutral conditions an extended model of vertical profiles of wind is the known logarithm profile: $U = \frac{u_*}{k}\ln\left(\frac{z}{z_0}\right)$, where k is the Von Karman constant ($k \approx 0.4$) and z_0 is called the roughness length depending on the terrain. In the case of this study a very high rate of measurement of three components of velocity is available; therefore vertical flux of momentum is utilized here.

In this section we study as the Fractal Dimension behaves versus the friction velocity. The friction velocity u_* presents maxima at day time closed to noon in every levels studied (5.8 and 13m) and minima at night. This maxima and minima are in concordance with the variation of FD in an analogous way to TKE treated in the last section (see Fig 13 a, c). The scatter plot for FD for w' component at the same levels and u· is showing in Fig 13b and d; it indicates an acceptable correlation between them. FD increase according to growing of u·. Again, a linear regression has been tried in order to quantify this correlation. Points in Fig 14 correspond to average values of FD of vertical components fluctuations against friction velocity. The positive slope and a good coefficient of correlation ($R = 0.961$) are outstanding results since they improve the ones in the TKE. This trend is observed in the u' and w' components at three heights studied.

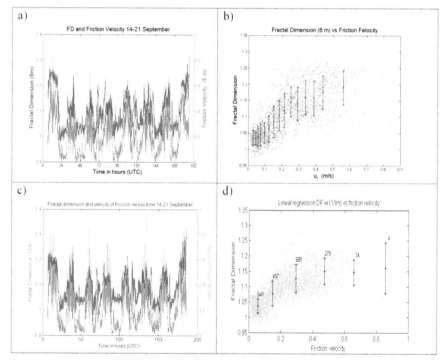

Figure 13. a) and c) Fractal dimension of fluctuation of w component at 5.8 m and 13 m (blue) and Friction Velocity at the same height (green). b and d) Fractal dimension versus Friction Velocity.

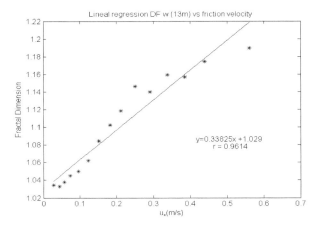

Figure 14. Linear regression of FD calculated for w' wind component versus friction velocity obtained from vertical fluxes of momentum (Eq. 6). The regression coefficients are presented (R = 0.9614).

5.4. Bulk Richardson number and fractal dimension

In this last Sub-section of results we try to investigate the relation between Fractal Dimension and the atmospheric stability in the Planetary Boundary Layer. We utilize the dynamic and thermal parameter denoted Bulk Richardson number, Ri_B, discussed in Section 4, and presented as a numerical approximation of the gradient Richardson number, Ri. The Ri_B value enables us to judge the atmospheric stability or, better named, the stability of stratification in the lower atmosphere. When Ri is small with any sign, the air flow is a pure shear flow driven by dynamic forces and we can say that layer is neutrally stratified. If Ri is near zero corresponding to neutrally stable boundary layer, one in which parcels displaced up and down adiabatically maintain exactly the same density as the surrounding air and thus experience no net buoyancy forces.

When Ri is large the air flow is driven by buoyancy. Positive values of Ri correspond with stable conditions and buoyancy forces keep out vertical displacement and mixing become less active. At the contrary, negatives and large Ri is in instability, where the thermal effects are doing air move vertically and fluctuations of wind components happen. Under instability turbulence increase and mixing of atmospheric properties: momentum, energy and mass (concentration of the components) are more efficient.

Thus, we go to exam whether the parameter Fractal Dimension can be an appropriated index to classify the atmospheric stratification. As it is already explained the numerical approximation used here is the Ri_B (Eq. 4). The main difficulty in the use of this parameter is the fact of which is not robust, since small shear in the wind produce values extremes of Ri_B, both positive and negative. In order to avoid such situations it has been remove of the study cases with $Ri_B < -5$ (convective) and $Ri_B > 1.5$ (strong stability, hard inversion). The time series of Ri_B along the complete period of study is drawn in Figure 15a. Can be see how negative values of Ri_B are in day time and positive or near-zero are far away of the central day.

The Fractal Dimension *versus* Bulk Richardson number shows a different behavior depending on the kind of stability as is shown in the Fig. 15b. That is, in strong instability it is observed mean values of the Fractal Dimension almost constant, around 1.15, but increasing when we are going to $Ri_B = 0$ (near neutral stratification). In neutral conditions the Fractal Dimension is maximum and decreases quickly for stable conditions. This result agrees with the relationships found for the potential temperature differences in the Fig. 7 and the temperature differences in the layer 50 – 0.22 m (Fig. 5). This behavior is similar at three heights studied.

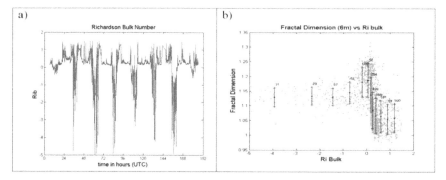

Figure 15. a) Bulk Richardson Number evaluated from the 32 - 5.8 m layer, along the whole period of study, where extreme values have been removed (-5< Ri_B< 1.5). b) Scatter plot of Fractal Dimension versus Bulk Richardson number; the maxima values of FD are in positive of Ri_B close to zero, and minima correspond to strong stability.

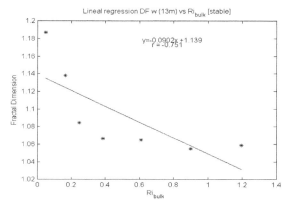

Figure 16. Linear regression of a set points, average FD in Ri_B intervals for the stable regime at height 13 m.

Behavior of FD according stability must be carry out separately in both kind of stability, as it was say before, instability give near constant values of Fractal Dimension except in quasi-neutrality, but in stability conditions, variation of FD it is clearly observed. In Fig. 16 it is shown the linear regression of mean values of the set points in stable stratification of the scatter plot of the Fig 13. As a result of that fit, can be conclude FD decrease as stability increase, note the negative slope, but the coefficient of correlation is weaker than in dynamic analysis cases (r = 0.751).

Maxima values of FD in near-neutral stratification it is supported for the results in the study of TKE or friction velocity, where enhance of dynamical effects gave a rising of the Fractal Dimension.

6. Conclusions

This Chapter has treated about the influence of the atmospheric conditions, dynamics and thermal, over the fractal structure of the wind near the ground. The results that have been obtained in the work presented along the Chapter, lead to main following conclusions:

- An easy method for obtain the Fractal Dimension of velocity component fluctuation in the PBL has been carried out. It has been based in the Kolmogorov capacity or box-counting dimension concept.
- In the 5.8 - 32 m layer studied the Fractal Dimension is greater when the differences of potential temperature are negative (instable), reaching maxima values at differences near to zero (near-neutral stratification), and with positive values (stable stratum) the Fractal Dimension is lightly inferior. This result is according to behavior of the thermal inversion in the stratum 0.22 - 50 m.
- There exist an increasing of the Fractal Dimension there exists with the growing of the two dynamic magnitudes studied: Turbulent Kinetic Energy and friction velocity. The behavior of TKE and friction velocity are similar at three heights analyzed. The values of Fractal Dimension with these dynamics parameters are maxima at day time, close to noon, and minima at night according to the turbulence variation in the daily cycle.
- The Fractal Dimension is depending on kind of stratification. For negative values of Bulk Richardson number FD keeps approximately constant but in stability FD decrease quickly with Ri_B. An acceptable correlation between FD and Bulk Richardson Number has been observed for positives Ri_B. In the neutral conditions the Fractal Dimension reach its maximum.

Finally, it can conclude that dynamical origin of the turbulence has a more clear relation than the thermal origin with the fractal structure of the wind, but both are important.

Author details

Manuel Tijera
Department of Applied Mathematics (Biomathematics), University Complutense of Madrid, Spain

Gregorio Maqueda* and José L. Cano
*Department of Earth Physics, Astronomy and Astrophysics II.,
University Complutense of Madrid, Spain*

Carlos Yagüe
Department of Geophysics and Meteorology, University Complutense of Madrid, Spain

---

* Corresponding Author

Acknowledgement

This research has been funded by the Spanish Ministry of Science and Innovation (projects CGL2009-12797-C03-03). The GR35/10 program (supported by Banco Santander and UCM) has also partially financed this work through the Research Group "Micrometeorology and Climate Variability" (No 910437). Thanks to teams participating in SABLES-98 for the facilities with the data.

7. References

[1] Frisch, U. (1995). *Turbulence.* Cambridge Univesity Press, England, 296 pp.

[2] Falkovich G. and Sreenivasan K. R. (2006). Lessons from Hydrodynamic Turbulence. Physics Today 59: 43-49.

[3] Falconer K. J. (1990). *Fractal Geometry Mathematical Foundations and Applications.* John Wiley & Sons, Ltd. 288 pp

[4] Peitgen H., Jürgens H. and Saupe D. (2004): *Chaos and Fractals.* Springer-Verlag 971pp.

[5] Grassberger P. and Procaccia I. (1983). Characterization of Strange Attractors. Phys. Rev Lett. 50, 346-349

[6] Abarbanel H. D. I., Rabinovich M. I., and Sushchik M. M. (1993). *Introduction to Nonlinear Dinamic for Physicists.* Word Scientific. 158 pp

[7] Kolmogorov A. N. (1941a). The local structure of turbulence in incompressible viscous fluid for very large Reynolds number, Dolk. Akad. Nauk SSSR 30, 299-303 (Reprinted in Proc. R. Soc. Lond. A 434, 9-13 (1991))

[8] Kolmogorov A. N. (1941b). On degeneration (decay) of isotropic turbulence in an incompressible viscous liquid, Dolk. Akad. Nauk SSSR 31, 538-540.

[9] Cuxart J., C. Yagüe, G. Morales, E. Terradellas., J. Orbe, J. Calvo, A. Fernández, M.R. Soler, C. Infante, P. Buenestado, A. Espinalt, H.E. Joergensen, J.M. Rees, J. Vilà, J.M. Redondo, I.R.Cantalapiedra, and Conangla, L. (2000). Stable atmospheric boundary layer experiment in Spain (SABLES 98): A report. Bound.-Layer Meteorol., 96, 337-370.

[10] Kaimal, J. C. and Finnigan J. J. (1994). *Atmospheric Boundary Layer Flows. Their Structure and Measurement,* Oxford University Press. 289 pp

[11] Vassilicos J. C. and J. C. R. Hunt, (1991). Fractal dimensions and spectra of interfaces with application to turbulence. Proc. R. Soc. London Ser. A 435-505.

[12] Hausdorff F. (1918). Dimension und äusseres Mass. Mathematische Ann. 79 157-179

[13] Tijera, M., Cano J.L., Cano, D., Bolster, B. and Redondo J.M. (2008). Filtered deterministic waves and analysis of the fractal dimension of the components of the wind velocity. Nuovo Cimento C. Geophysics and Space Physics 31, 653-667

[14] Stull R. B. (1988). *An introduction to Boundary Layer Meteorology.* Kluwer Academic Publishers.

[15] Arya S. P. (2001). *Introduction to Micrometeorology.* International Geophysics Series, Academic Press. 420 pp

[16] Lynch A.H and Cassano J.J. (2006). *Atmospheric Dynamics.* John Wiley & Sons, Ltd. 280 pp

[17] Garratt, J. R. (1992). *The atmospheric boundary layer.* Cambridge University Press, 316 pp.

Chapter 4

Evolution of Cosmic System

Noboru Tanizuka

Additional information is available at the end of the chapter

1. Introduction

What does it mean that the cosmic radio wave flux density varies with the passage of time is an interesting question; the radio wave is of the quasar, a system of galaxy, which is distributed in our universe from a few billions of light years to the distanse close to the big bang age and has been radiating immense electromagnetic energy from it by the synchrotron radiation that we may able to make a measurement of the flux density at micro wave bands with a radio interferometer[3,4]. A group of radio observers and astronomers has been monitoring daily so far over several years extragalactic radio sources (radio galaxies, quasars, etc.) and the monitored data were kindly shared with us who were interested in using for analysis[5]. In a few recent decades, the chaos and fractal theory has been intensively studied and developed in the fields of mathematics, computer numerical analysis, natural sciences and technologies[1], and in same decades, the nonlinear time series analysis methods have been developed intensely based on the newly understood ideas of the theory for analyzing the nonlinear phenonena[2].

The study in this chapter is motivated by the three factors mentioned above to analyze the time series of the radio wave flux density from the cosmological object, primarily, with one of the nonlinear methods, for finding the dynamics related to the cosmic object, including its information in the flux density variations. We hoped that if we could infer the dynamics and if the result would be found to have any rule changing with the magnitude of the red shift of the object we might have some knowledge concerning to the evolution of our universe. The hope has been prompted us to continue consistently to analyze the time series data. The period of monitor over several years is extremely short compared with the cosmic age, however , the analysis result of the time series data in newly developing methods may give us a new sight viewed from the nonlinear dynamics in the short time scale for the cosmic dynamical system.

2. Linear and nonlinear systems

2.1. Linear system

The result of a linear data computed with Fourier spectral analysis gives a pure periodicity, which means predictability of event(s) for unlimited future. No event of evolution can be expected by a linear data, and a linear system is not a way of our natural world. On the other hand, the result of a pure random data computed with it gives no structure of the periodicity, a continuous spectrum, and means no event of predictability.

2.2. Nonlinear system

One of discrete nonlinear dynamical systems is given in this section for explanation. The logistic map function is given in Eq.(1) with a difference equation. Given that the time series data is generated from this system by iterating at $t = 0,1,..,N$.

$$f(x) = ax(1 - x); \; x(t + 1) = f(x(t)) \tag{1}$$

The result of the Fourier analysis for the time series data at parameter $a = 4.0$ in the function of Eq.(1) is indistinguishable from the result of the random data. The spectra for the random data and logistic time series data are given in Fig.1. The latter comes from the deterministic system, but the analysis result does not disclose the nonlinearity of the data. This is a reason why a help of a new method is necessary for the analysis of data from the natural phenomena which are often complex for us to understand and may include such complexity as above in part. A method to distinguish between the kinds of data is to draw a graph composed of coordinates with time-lagged components, that is, a graphical execution of differentiation for a time series. The result of the graphical execution for the random and logistic time series in Fig. 1 is shown in Fig. 2. As a result, in the figure the two time series are correctly discriminated. The time series generated from Eq.(1) resulted in a quadratic expression with the second dimension of coordinates (see Fig. 2(b)). On the other hand, the random time series resulted in the uncorrelated graph as shown in Fig. 2(a).[1]

The time series data measured from natural and social systems is not generated in such a mathematical way and so complicated that more intricated process is required to deal with data. This problem will be discussed in later sections. Before we do this, some methods of quantifying fractal data and some characteristics of chaos dynamics (the initial value dependence and the parametric dependence) are exemplified by the logistic system.

Time series data were generated from Eq.(1) at parameter $a = 3.7$ with two initial values at $x(0) = 0.10000$ and at $x(0) = 0.10001$, and the two time series are drawn as the diagram shown in Fig. 3 (left). The diagram shows that the trajectories of the two time series separate in a few tens of iteration time, which means that a small error was extended to a magnitude of space the time series occupies within the limited time and the prediction is failed over the

---

[1] It is true in the second dimension of coordinates. If the time series is generated by the function of higher degree, the correlation may be true in the graph at a higher dimension of coordinates.

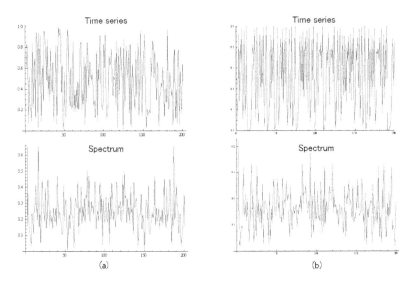

Figure 1. Fourier analysis for time series data of (a) uniform random numbers and (b) logistic map system at $a = 4.0$. The spectrum is shown by the absolute of the Fourier transform.

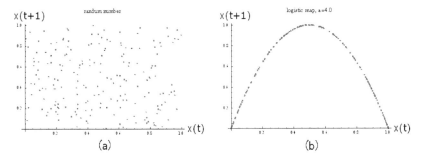

Figure 2. Graphical analysis for (a) random time series data and (b) logistic time series data given in Fig. 1.

time. This means that in spite of extremely small error in the chaos deterministic system, the error is extended to a scale of state space in a limited time, that the system is unpredictable. The fact was explained in the Fourier spectrum of the time series in Fig. 1 (b).

The parametric dependence is characteristic to a chaos system. The vertical axis on the right diagram in Fig. 3 shows the values of number computed by Eq.(1) at parameter **a** in the range of $3.8 \le a \le 4.0$, along the horizontal axis. At a parameter value between 3.82 and 3.83, the behavior of the system drastically changes to periodicity. In the logistic system the parameter is interpreted as the environment for a living thing to survive. For nonlinear systems, a parameter value becomes crucial for the system's behavior. For example, in the logistic dynamics, extinction or evolution of the system depends on the parameter value.

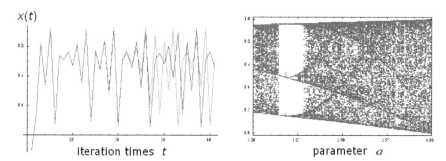

Figure 3. Logistic time series at parameter $a = 3.7$ and at initial values $x(0) = 0.10000$ and $x(0) = 0.10001$ (orange) on the left, and the bifurcation diagram in the parameter range over $3.8 \leq a \leq 4.0$ on the right.

3. Quantitative properties of nonlinearity

Time series data of a natural system may often be of a nonlinear dynamics, about which we know little for the system and need to analyze in an appropreate method in assumed state space dimensions. In this section we discuss on the way how to quantify such data assumed to be nonlinear by exemplifying the analysis for the logistic system.

3.1. Lyapunov exponent and information entropy

The initial value dependence of a system is evaluated by the Lyapunov exponent given by

$$\lambda = \lim_{N \to \infty} \frac{1}{N} \sum_{t=0}^{N-1} \log|f'(x(t))| , \tag{2}$$

for the discrete system, and $\lambda = \lim_{t \to \infty} t^{-1} \log d_t / d_0$ for the continuous system with d_0, d_t the initial error and its expansion at time t, respectively. It is easy to understand that if the system is in the chaos, $\lambda > 0$, the error is exponentially extended. Even if we have this way to distinguish a system whether it is a chaos system or a mere random system, it is a difficult problem to analyze λ of a time series data because the system's function f is not in our hand.

The information entropy of a system, as its manifold is given in a state space, is defined as

$$H(\varepsilon) = - \sum_{i=1}^{M} p_i \log p_i , \quad \sum_{i=1}^{M} p_i = 1 , \tag{3}$$

where ε is an infinitesimal length with a super cube and M the number of cubes with which cover whole manifold in the state space. Equation (3) quantifies the distribution of data points in the state space as the average number of the amount of information.

Figure 4 shows the bifurcation diagram for the logistic system (left diagram), the Lyapunov exponent λ (middle diag.) and the information entropy $H(\varepsilon)$ (right diag.), with common abscissas of parameter a ($2.8 \leq a \leq 4.0$). It is clear that the bifurcation diagram (left diagram) is quantified by the Lyapunov exponent (middle diag.: not chaos in $\lambda \leq 0$; chaos

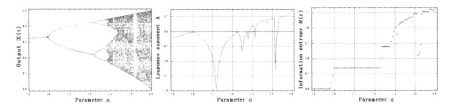

Figure 4. The bifurcation diagram of the logistic map (left diagram) over the parameter range $2.8 \leq a \leq 4.0$, the Lyapunov exponent (middle diag.) and the information entropy (right diag.).

in $\lambda > 0$) and by the information entropy varieing with parameter a (right diag., compare with the left diag.).

3.2. Fractal dimension

In this chapter we aim to infer the cosmic system's evolution by the flux density data radiated from cosmic object and measured by the interferometer (of the radio wave). The fractal dimension is useful to study the system with which the data is related. We have a variety of fractal dimensions; the box counting dimension D_0, the information dimension D_1 and the correlation dimension D_2 as defined in the following equations [see Fig. 5.].

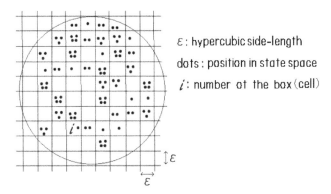

ε : hypercubic side-length

dots : position in state space

i : number of the box (cell)

Figure 5. Diagram of a point distribution in a state space for computing the fractal dimension.

$$D_0 = \lim_{\varepsilon \to 0} \frac{\log M(\varepsilon)}{-\log \varepsilon} \, , D_1 = \lim_{\varepsilon \to 0} \frac{H(\varepsilon)}{-\log \varepsilon} \, , \ D_2 = \lim_{\varepsilon \to 0} \frac{\log \sum_{i=1}^{M} p_i^2}{\log \varepsilon} \tag{4}$$

The fractal dimensions in Eq.(4) are derived from the generalized dimension D_q:

$$D_q = \lim_{\varepsilon \to 0} \frac{1}{q-1} \frac{\log \sum_{i=1}^{M} p_i^q}{\log \varepsilon} \tag{5}$$

for $q = 0$ (the box counting dimension), $q = 1$ (the information dimension at $q \to 1$) and $q = 2$ (the correlation dimension) [1]. It is not easy to compute the fractal dimension from measured data, even if it is at $q = 0$, because we know little about the system's dynamics, on which we are to study, and in addition, it is impossible to box-count a manifold constructed in a state space with unknown dimension. Fortunately, we have Grassbergar-Procaccia algorithm (GPA) to compute the measured data and substitute the result for the fractal dimension at $q = 2$ [2]. The GPA is discussed in section 5.

4. Dynamical system and time series data

The purpose of this section is to discuss on the problem: if we measure the time series data from a system of nature how we access to a function of the system with which the data is generated. If the nature was constructed by the mathematics it would have been going well to solve the problem. Unfortunately, the nature, I believe, do not go so easy. We need to solve it by devising the data reconstruction and by applying above nonlinear methods to it .

4.1. Time series data

The originally measured data is defined as follows: A measurement starts at time t_0 with sampling rate τ and the time series is expressed in the following way, with i natural number

$$y(t_0), y(t_0 + \tau), \cdots, y(t_0 + i\tau), \cdots$$

If we find a time lag τ' proper for the system in three dimensional state space, the time series is reconstructed, for example, to three dimensional vector $\mathbf{V}_t$ at time t.

$$V_t = (y(t), y(t + \tau'), y(t + 2\tau')); \; t = 1,2, \cdots$$

Vector $\boldsymbol{V}_t$ is embedded in three dimensional time-lagged state space. As the time goes by, the vector draws a trajectory in the state space. The measured data may geometrically express its functional property in this way. The method can be considered to differentiate the data in the state space in a graphycal way. The dimension of the system is unknown in advance, so an original dimension of the reconstructed vector must be searched by changing it one by one until to find the optimal one which is called the embedding dimension. The dynamics of the observed system have a fractal dimension in the embedding dimension. A manifold is drawn in the embedding dimension in this way from the ofserved data and it is called the attractor of the system given by the observed data. From the manifold we infer the original function, as a mathematical nonlinear equation.

4.2. Embedding theory

The attractor is the manifold of a dynamical system, from which a physical quantity is continuausly released to be observed and the time series is, as a result, accessed to be analized. The time series is reconstructed in the form of a vector V_t at m-th dimension to be embedded in the state space. In the following equation the time lag τ' was replaced by τ.

$$V_t = (y(t), y(t + \tau), \cdots, y(t + (m - 1)\tau)) \tag{6}$$

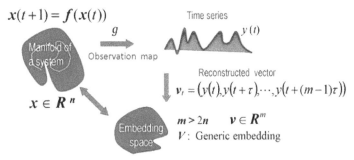

Figure 6. Takens' theorem given by a schemetic diagram to infer the system's dynamics from measured time series[2].

The data set $\{y(t)\}_{t=t_0}^{t_N}$ is measured by gM, with g the observation map and M the manifold of the system's (source) dynamics at n-th dimension. Takens' theory claims that the attractor reconstructed on the embedding space in Fig. 6 is generic embedding under condition $m > 2n$.[2] The attractor in the embedding state space is a theoretical reflection of the manifold, that is, the embedding map is $V: M \to R^m$, in which the condition does not need to be satisfied. The source dynamics at the observed system could be inferred in this way. It is generally impossible to solve the function, in a definitive form, of the system's dynamics.

Takens' theorem is summarized in Eq. (7) ~ Eq. (9). h is a map function transforming the embedded attractor into the original attractor.

$$f: M \to M \quad \Rightarrow \quad g: M \to R \quad \Rightarrow \quad V: M \to R^m \tag{7}$$

$$V(x) = \left(g(x), g(f(x)), \cdots, g(f^{m-1}(x)) \right) \tag{8}$$

$$h: V \to M \tag{9}$$

The experimantal expression of Eq.(6), reconstructed by using measured data, corresponds to the theoretical expression of Eq.(8), assuming that a time delay is neglected and the time span is same between both systems. We have the attractor V by analyzing measured data, but it is difficult to have a deterministic expression of the manifold M in Eq.(9).

It is same to say that Galileo Galilei could find experimentally the gravity on the earth, but could not express it in a deterministic expression as the Newton's equation. The map (h) in Eq.(9) is similar to the gravity in the era. At present, it is only possible to get access to the geometrical manifold with the way given in this section. A discussion on the method how to apply the fractal dimension to quantify the geometrical manifold will be given.

It is useful to give attention to noise inevitaby coming into the dynamical system and the observation system. The noise comes into the two systems[2],

$$x(t + 1) = f(x(t)) + \eta(t) \tag{10}$$

$$y(t) = g(x(t)) + \xi(t) \tag{11}$$

where $\eta(t)$ and $\xi(t)$ are system's noise and observation noise, respectively. This makes it very difficult to judge observed data that system's function f is a genuine chaos or not, even if the analysis gives a chaotic result, because the chaos trajectory depends severely on the initial condition.

4.3. Correlation dimension

The attractor is reconstructed in the embedded state space from observed time series. We are interested in the dynamical system of the quasar, the galactic object distributed at cosmological distances (up to ten billions of light years), which releases vast energy by synchrotron radiation and enables us to research some cosmic information by the fluctuations of the microwave flux density. We study the dynamics involving in the time series to know the structure of the dynamics for the system at different cosmological distances. We may be to have some dynamical knowledge of the systems's evolution in the experimental method. The system's information is, in our context, is the manifold for the dynamics of the quasar system at different cosmological distances. The data is the time series of the flux density of the microwaves, 2.7GHz and 8.1GHz, for more than twenty quasars, daily monitored over thousand days. The fractal dimension of the manifold is to be analyzed. The dimensions introduced in the section 3 are difficult to compute with the reconstructed data. Fortunately we have an useful method to compute the correlation dimension developed by Grassberger and Procaccia (GPA) as the substitute of D_2. The algorithm for calculation are the correlation sum and the fractal exponent in the following expressions, N the number of points in the embedding state space at m-th dimension.

Given each point j, k ($j \neq k$) in Fig. 5 (not number of the cell), with N the number of full points and r the diameter of hypersphere, the correlation sum is expressed by [1,2]

$$C^{(m)}(r) = \frac{1}{N} \sum_{k=1}^{N} \frac{1}{N-1} \sum_{j=1}^{N-1} \theta[r - \|v(j) - v(k)\|], \tag{12}$$

in which $\theta[\cdot]$ is the Heaviside function, counting a pair of m-th dimensional vector points whose distance is within r over all pairs of points. Eq.(12) counts the probability that any pair of state space points meets within a length of r in m-th dimensional state space. The fractal dimension at m-th dimensional state space is expressed by (see Eq.(4))

$$D_2^{(m)} = \lim_{r \to 0} \frac{\log C^{(m)}(r)}{\log r} \tag{13}$$

The optimal correlation dimension D_2 for the attractor embedded in the optimal state space for the time series in question is defined as $D_2^{(m)}$ at $m = m_s$ when the value of $D_2^{(m)}$ ceases to increase as the increase of m. This means that at a full embedding dimension, any points in state space of the attractor can not occupy same position by the no-intersection theorem of chaos [1]. The D_2 is a fractional dimension and the m_s is the embedding dimension in which the system works ($m_s - 1 < D_2 < m_s$) .

4.4. Other methods of analysis

We introduce briefly three methods of analyzing the characteristics of the flux density variation, in which the same data were also computed for reference. The results will cover different aspects of the variation. The first is the spectral index α : The modulus of the fourier spectrum $|A|$, the intensity of the fluctuations at frequency f is computed over the period of 1024 days; the method of least squares was used for the pairs of the logarithms of $|A|$ and f in the frequency range of $\frac{1}{100} \leq f \leq \frac{1}{10}\frac{1}{day}$ to eliminate the major error outside the frequency range. The power law $|A| \propto f^{-\alpha}$ was computed for all of the observed data. The second is the Higuchi's fractal dimension D [9]. The absolute change of the flux density at interval k averaged over the period (1024 days) is computed. Given the averaged change L, the relationship $L \propto k^D$ stands for the interval $1 \leq k \leq 100$ days. The detailed algorithm can be reffered in reference [9]. The dimension D expresses a complexity of the variation in the range $1 \leq D \leq 2$, that is, from liner, $D = 1$, to the plane, $D = 2$, from fractal dimensional view. The third is the Hurst exponent H [10]. Given the ratio R/σ over the interval τ, with range R a difference from top to bottom levels of the reconstructed time series $Y(t) = \sum_{i=1}^{t}(y(i) - \bar{y})$, $\bar{y}$ and σ the average and the standard deviation of $y(i)$, respectively, in the interval τ. The relationship $R/\sigma \propto (\tau/2)^H$ stands in the range of $10 \leq \tau \leq 1000$ days.

The detailed introductions for the methods be referred in [9-11][2]. The result analyzed in these methods will be shown later. The results computed in above principles are useful for cross-checking the knowledge of the result of source dynamics.

5. Time series data

The extragaractic radio sources generate the time series data of the radio wave flux density for us to observe and to analyze their system's dynamics to see a mechanism how the cosmic object has been evolved in the cosmological age from a dynamic aspect of view.[3]

5.1. Monitored cosmic objects

Compact extragaractic radio sources had been monitored daily by Waltmann et al. at GBI radio wave observatory over 3000 days from 1979 [5,6]. Waltmann et al were kind to send us the data of 46 extragalactic objects, from which 21 QSOs and 7 BL Lacs were selected for analysis. At the beginning we analyzed the data in the methods of the spectral index, of the Higuchi's fractal dimension and of the Hurst exponent. The methods will be explained briefly, and the result was published in [7].

The monitored microwave frequencies were at 2.7 and 8.1GHz; and the red shift (the indicator of cosmological distance) of the monitored objects ranged from 0.15 to 2.22 (from one billion to ten billions of light years). The name of the objects are shown in Table 1 as

---

[2] $1/f$ noise characterized as the power law events in the electronic circuit is in reference [11].

well as the red shifts. In Figure 7, the diagrams of the flux density variation over nine years are shown for several quasars.

0133+476*	0202+319	0224+671	0235+164*	0237-234	0333+321	0336-019
0.860	1.466	0.524	0.851	2.224	1.253	0.852
0420-014	0552+398	0828+493	0851+202*	0923+392	0954+658	1245-197
0.915	2.365	0.548	0.306	0.699	0.368	1.275
1328+254	1328+307	1502+106	1555+001	1611+343	1641+399	1741-038
1.055	0.849	1.833	1.770	1.404	0.595	1.054
1749+096*	1749+701*	1821+107	2134+004	2200+420*	2234+282	2251+158*
0.322	0.760	1.036	1.936	0.070	0.795	0.859
Symbol *	BL Lac (7)	; QSO (21)			;	

Table 1. The Name of the extragalactic objects selected for analysis and the red shift [7,8]. Data used for analysis is of 1024 days (start from 2 Feb.1984 observed on daily basis.

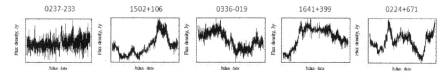

Figure 7. Radio wave flux density variation at 8.1GHz monitored daily over three thousand days. Panels listed in descending order of the magnitude of the red shift. Abscissa: Julian date from 44000 to 47000; ordinate: flux density Jy (10^{-28}W/m^2/Hz) from a few Jy to ten Jy.

5.2. Process of analysis

5.2.1. Correlation dimension

The method of estimating the correlation dimension of the flux density time series of the quasars listed in Table 1 is plainly described here. The correlation sum for the time series data is calculated in the method given in Eq.(12). The time series for the quasar 0224+671, for example, is reconstructed in the way given in subsection 4.2 and embedded in the reconstructed state space (at $m = 3$, for example) as shown at the left diagram (state space diagram) in Fig. 8. The correlation sum is plotted in log scales of abscissa and ordinate according to the diameter r changing its size step by step at each dimension m as shown at the middle diagram in the figure. The correlation exponent is estimeted from the inclination of the correlation sum graph in the way described in Fig. 8. The correlation exponent increase as the increase of the embedding dimension, as shown at the right diagram in Fig. 8, up to the state space dimension to become enough for the original attractor to be contained in the state space. If the exponent stops to increase at an embedding dimension, the exponent, a magnitude less than the embedding dimension, is called the correlation dimension, the fractal dimension, which is thought to reflect the active number of the variables with which the object's system works. It should be taken care that the reliability of the estimated fractal dimension is limited by the number of the data [12]. In our case, the dimension seems to be appropriate values and useful because it will bring cosmological information on the dynamics of object system which is expected to vary with the red shift.

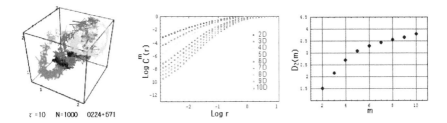

$\tau = 10$ $N = 1000$ $0224 + 571$

Figure 8. Left diagram: Attractor of the flux density time series for quasar 0224+571 embedded in the reconstructed state space at $m = 3$ ($\tau = 3, N = 1000$), using different colours at time span clarifying the change of state with the passage of time. Middle diagram: Correlation sum computed by using Eq.(12) for each embedding dimension m (parameter from 2 to 10). Right diagram: The correlation exponent was estimated by the inclination of the each curve in the middle diagram at the first order fitting of the least square in the range -3.0<Log r<-1.5. The correlation exponent increases with the increase of the dimension m.

Other indices

A process of analyzing the data with other methods introduced in subsection 4.4 is shown in Fig.9. The diagrams gives us an insight of the way how each index is derived. We will show all of the result analyzed in these methods for a cross reference with the correlation dimension.

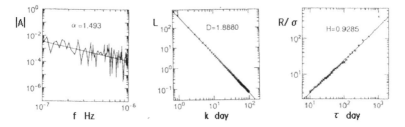

Figure 9. A computing process of analyzing other indices as an example for the time series of the radio wave at 8.1GHz of QSO 1641+399. Left diagram : Spectral index, Middle diagram : Higuchi's fractal dimension, Right diagram : Hurst exponent.

6. Result of analysis

6.1. Correlation dimension

The correlation dimension reflects the dimension of a dynamical function $f(x)$, which is thought to relate closely to the dynamics of monitored radio source. The observation map function g includes the path of radio wave p of the cosmic space and the observation system (antenna) g'; consequently $g = g'p$ stands for the observation map. In Fig.10 we show the analysis result of the correlation dimension versus the red shift of the object for the data given in Table 1. The result shows a tendency for the both microwaves (2.7 and 8.1GHz) that the correlation dimension increases as the increase of red shift z. It could be said that the complexity of the system's dynamics is increasing as the distance to the quasar is farther. The cosmic spatial map p is unknown for us; on the other hand the data is accessed by $gM = g'pM$, then the system's complexity can be said to depend on the path (map p). In this point, as far as I learned from a radio astronomer at National Astronomical Observatory (Japan), his view was that the influence of p must be weak; if we admit this view, the order in the source dynamics becomes less as the cosmological distance is closer to the Big Bang because the correlation dimension of the system's dynamics can be considered to be a complexity of the system's behavior.

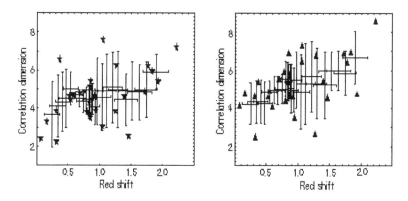

Figure 10. Correlation dimension of the reconstructed attractor for the flux density time series of the quasars given in Table 1 versus the quasar's red shift. Left diagram : 2.7 GH, Right diagram : 8.1 GHz. Horizontal and vertical bars : Running means with error bars at 5 steps along both axes (the red shift and the correlation dimension).

It may be taken care to see the diagram in Fig.10 that the radio wave frequency (2.7 or 8.1 GHz) from which the correlation dimension was derived is the value on the earth; the frequency at the radio wave source must be modified by the red shift (See Table 2); the second is that the sampling rate (one day on the earth) must be also modified on the quasar by the theory of relativity (See Table 2). [7]

red shift z	0.1	0.5	1.0	1.5	2.0	2.5
$1 + z$	1.1	1.5	2.0	2.5	3.0	3.5
$\sqrt{1 - \beta^2}$	0.99	0.92	0.80	0.68	0.60	0.52

Table 2. The Doppler and the relativistic effects. The radio wave frequency is multiplied by $1 + z$ on a quasar at z and the sampling rate on it is multiplied by $\sqrt{1 - \beta^2}$, where β is the ratio of the recession velocity to light velocity.

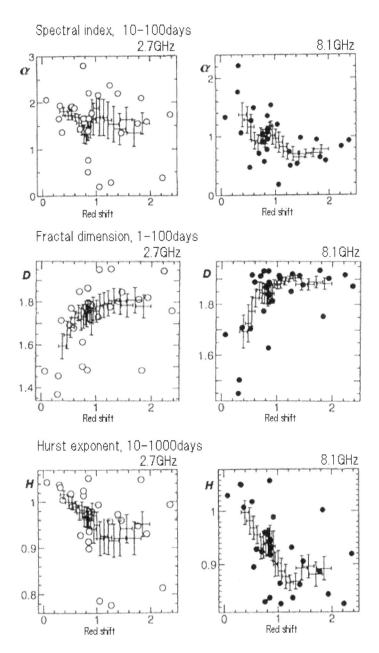

Figure 11. The indices of the flux density time series versus the red shift [7].

6.2. Other indices

Figure 12 shows the result of the indices analyzed in the methods introduced in subsections 4.4 and subsubsection 5.2.2: the diagrams of the spectral index α, the Higuchi's fractal dimension D and the Hurst exponent H versus the red shift z with the holizontal and vertical error bars by calculating the running means over seven points. It may be clear that the indices, α, D and H, vary according to the increase of the red shift in the manner not inconsistent with the correlation dimension D_2 versus z (see Fig. 10). It must be taken in consideration that the indices have their each reflection to the characteristic period due to the algorithm of analysis. It is interesting to see in the graphs of the indices that a typical discontinuity is present at red shift close to $z \approx 1$ (α, H) and the indices do not vary beyond red shift 1.2, $z > 1.2$ (α, D, H). The indices relate to the complexity of how the flux density varies with time; the complexity, α in the frequency domain, D in the fractal dimension of the graph for the density variation and H in the trend persistency. It is useful for us in the empirical way to see the relationship among three indices in our case (see Fig. 12). It may be interesting to see that the relationship between H and D has a systematic dependence despite of the different period concerned, in which the index is based, ten times as much (see Fig.12).

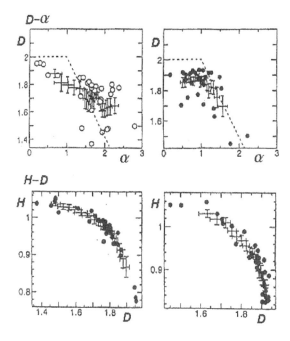

Figure 12. The relationship between the indices D and α (top), and between the indices H and D (bottom) for the 2.7 GHz (left column) and the 8.1 GHz (right column) [7].

6.3. Incident angles of the radio wave to the solar system and to our galaxy

It is natural to have a question that the radio wave flux may be strongly scattered by the matters whose density may be high around the earth (our solar system) and our galaxy; if it is true, there may be a possibility that the indices may affected by the insident angles to our solar system and to our galaxy. The distributions the index D versus right ascension, vs. declination, vs. galactic longitude and vs. galactic latitude are shown in Fig. 13 and Fig. 14. The distributions of the indices other than D may be inferred from the relationships given in Fig. 12. As shown at the bottom ranks in Fig. 13 and Fig. 14, the distribution of the red shift versus the insident angles is almost the same as the distribution of the index versus the insident angles. This became clear from the view in the moving average as give above. We infer that the flux density variation may not be caused by the matters around the earth (our solar system) nor our galaxy, but by any factors related with the red shift (the radio wave path as long as the cosmological distances) and by the system of the radio wave source, the quasar.[13]

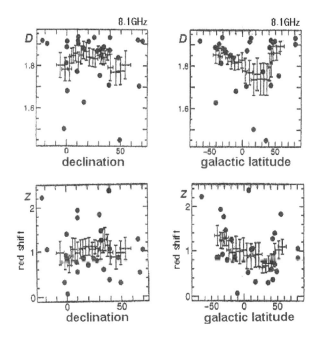

Figure 13. The distribution of the fractal dimension D versus the declination (top, left) and vs. the galactic latitude (top, right) for the radio wave at 8.1GHz, and the distribution of the red shift versus the declination (bottom, left) and vs. the galactic latitude (bottom, right). The vertical and horizontal error bars are of the running means for the numerical values computed by every seven steps. [7]

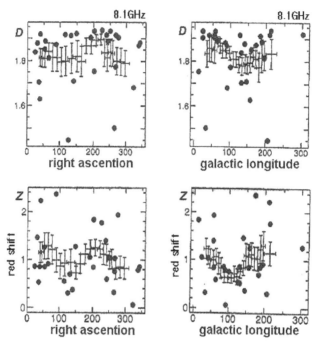

Figure 14. The distribution of the fractal dimension D versus the right ascention (top, left) and vs. the galactic longitude (top, right) for the radio wave at 8.1GHz , and the distribution of the red shift versus the right ascention (bottom, left) and vs. the galactic longitude (bottom, right). The vertical and horizontal error bars are of the running means computed by every seven steps.[7]

7. Conclusion

The correlation dimension D_2 of the flux density variation of the quasar radio wave increases, on average, with the increase in the red shift of quasar up to $z \approx 1$ and reaches a limit at $z \approx 1$, and the dimension depends mainly on the red shift and seems to be not affected, on average, by the incident angles to the earth and to our galaxy, from the view based on our analysis. The result is important because the numerical value of the correlation dimension includes the dynamical dimension of the flux density variation monitored over a thousand days, which reflects a source dynamics as considered in the theory given in section 4 and the dependence on the red shift does not conflict with the dependency of other indices, though we have not yet the knowledge of the external modulation, or the map p, transfering the source dynamics to the antenna. We could not help using such task because of the limitation of the data monitored unintended for our analysis. A systematic and designed observation will be needed to collect data for our analysis. If the analysis will be possible to make based on the purposed data, we will have more reliable reflection of the system's dynamics, $v_{t+1} = f_z(v_t)$, where f_z is the function at red shift z, in a form of the manifold but not of the equation, and infer a more exact view of the cosmological evolution.

For readers from different fields the literature on the radio galaxies and quasars may be referred in reference[3].

As discussed in section 2., the nonlinear dynamics is sensitive to the initial condition and the parameter. Our hope is to infer a definite dynamical function f_z at red sfift z. The difficulty comes from the following notion due to noise inevitable in the actual system; $\eta(t)$: noise added to the genuin dynamics, and $\xi(t)$: noise added to the observation system;[2]

$$x(t+1) = f_z(x(t)) + \eta(t) \tag{14}$$

$$y(t) = g(x(t)) + \xi(t) \tag{15}$$

We may need to take in mind that our accessed data might have added by noise of Eq.(15) and the analyzed attractor (a reflection of f_z) might have been added by noise of Eq.(14). After all a more complicated process might have been taken into account to estimate our result and study on the evolution of cosmic system.

Author details

Noboru Tanizuka
Complex Systems Laboratory, Tondabayashi, Japan

Acknowledgement

I would like to thank many Japanese principal radio astrophysicists for giving their knowledges on this field, comments and frequent encouragements in meetings, Dr. E.B. Waltmann and her group for sending their data to the computer center at Osaka Prefecture University, Mr. M. Takano and Dr. M.R. Khan for their computing works as graduate students at OPU.

8. References

[1] Hilborn RC. Chaos and Nonlinear Dynamics. Oxford: Oxford Univ. Press; 2000.

[2] Aihara K. Fundamentals and Applications for Chaos Time Series Analysis. Tokyo: Sangyo Tosho; 2000.

[3] Kellermann KI., Owen FN. Radio Galaxies and Quasars. In: Verschuur GL., Kellermann KI. (eds.) Galactic and Extragalactic Radio Astronomy. New York: Springer-Verlag; 1988, p563-602.

[4] Akabane K., Kaifu N., Tahara H. Cosmic Radio Astronomy. Tokyo: Kyoritsu; 1988.

[5] Waltmann EB., Fiedler RL., Johnston KJ., Spencer JH.,Florkowski DR., Josties FJ., McCarthy DD., Matsakis DN. Daily Observations of Compact Radio Sources at 2.7 and 8.1GHz: 1979-1987. Astrophys. J. Suppl. Ser. 1991;77(Nov) 379-404.

[6] http://www.gb.nrao.edu/fgdocs/gbi/gbint.html (accessed 14 May 2012)

[7] Tanizuka N., Takano M. Observational Study on a Process of Evolution of Galaxies. IEE J Trans. C 2000;120-C(8/9) 1149-1156.

[8] Tanizuka N., Khan MR. Knowledge from the Time Series of Quasar Radio-Wave Flux Density. Systems and Computers in Japan 2003; 34(10) 56-62.

[9] Higuchi T. Approach to an irregular time series on the basis of the fractal theory. Physica 1988; 31 (D)277-283.

[10] Feder J. Fractals. New York: Plenum; 1989.

[11] Gupta MS., editor. Electrical Noise: Fundamentals & Sources. New York: IEEE Press; 1977.

[12] Ruelle D. Deterministic Chaos: The Science and the Fiction. Proc. R. Soc. London 1990; 427 A: 241-248.

[13] Tanizuka N. Analysis of Quasar Radio Wave Flux Density Fluctuations and its Cosmological Meanings. In Macucci M, Basso G. (eds) Noise and Fluctuations, 20th International Conference on Noise and Fluctuations, ICNF2009, 14-19 June 2009, Pisa, Italy. Melville, New York, AIP Conference Proceedings 1129: 2009 .

Chapter 5

Dynamical Model for Evolution of Rock Massive State as a Response on a Changing of Stress-Deformed State

O.A. Khachay, A.Yu. Khachay and O.Yu. Khachay

Additional information is available at the end of the chapter

1. Introduction

Geological medium is an open dynamical system, which is influenced on different scales by natural and man-made impacts, which change the medium state and lead as a result to a complicated many ranked hierarchic evolution. That is the subject of geo synergetics. Paradigm of physical mesomechanics, which was advanced by academician Panin V. E. and his scientific school, which includes the synergetic approach is a constructive method for research and changing the state of heterogenic materials [1]. That result had been obtained on specimens of different materials. In our results of research of no stationary geological medium in a frame of natural experiments in real rock massifs, which are under high man-made influence it was shown, that the state dynamics can be revealed with use synergetics in hierarchic medium. Active and passive geophysical monitoring plays a very important role for research of the state of dynamical geological systems. It can be achieved by use electromagnetic and seismic fields. Our experience of that research showed the changing of the system state reveals on the space scales and times in the parameters, which are linked with the peculiarities of the medium of the second or higher ranks [2 – 5].

It is known that the most geological systems are open and non equilibrium, which can long exist only in the regime of energy through circulation. The closing of the energy flow leads to the system transfer to a conservation stage, when the duration of its existence depends on its energy potential due to accumulated energy on the previous stage [4]. On a certain stage of open dynamical system evolution, exchanging by matter and energy with the surrounding medium, decays on a set of subsystems, which in their turn can decay on smaller systems. The criterion of defining boundaries of these systems is one of synergetic law: macroscopic processes in the systems, which exist in a non linear area with a self organization processes, are performed cooperative, coordinated and coherent. The base of the processes of self organization in the open non equilibrium geological systems is the energetic origin. If the energy potential does not achieve its threshold value, the processes of self organization do not begin, if it is sufficient for compensate losses to the outer medium, in the system will begin the processes of self organization and form space-time or time structures. The transition from chaos to a structure is performed by a jump. If the income of the energy is too much, the structurization of the medium finishes and the transition to the chaos begins.

In arbitrary open dissipative and nonlinear systems are generated self-oscillating processes, which are sustained by outer energy sources, due to self organization exists [4]. The research of the state dynamics, its structure and effects of self organization in the massif we can provide with geophysical methods, set on a many ranked hierarchic non stationary medium model.

2. Physical models and mathematical methods of research

From the mathematical point of view dynamical system is an object or process, for which the concept of a state is defined as a set of values in a given time and an operator, which defines the evolution of the initial state in time [6]. If for the description of system state evolution it is sufficient to know its state in a given moments of time, that system is denoted as a system with discrete time. Let the set of numbers is defined as $x=\{x_1, x_2,.. x_N\}$ in a some time moment describes the state of a dynamical system and to different sets $\{x_1, x_2,.. x_N\}$ correspond different states. Let us define the evolution operator, indicating the velocity of changing of each system state as:

$$\frac{\partial x_i}{\partial t} = F_i(t, x_1, x_2, ..., x_N), i = 1, ...N \qquad (1)$$

x – point of Euclid space $\mathcal{R}_N$, which is named as phase space, x – phase point. The system (1), for which the right part does not depend from time, is named autonomous. By research of dynamical system, which describes the change of oil layer state by vibration action, the right parts of equations (1) will depend from time, and the system will not be autonomous. If the system (1) complete with initial conditions x (0) =x0, we shall obtain an initial conditions problem (problem Koshi) for (1). The solution $\{x\ (t), t>0\}$, which belongs to a set of points of phase space $\mathcal{R}_N$, which forms a phase trajectory; vector-function F(x) specifies the vector field of velocities. The phase trajectories and vector field of velocities give a descriptive representation about the system behavior character during time. The set of phase trajectories, which correspond to different origin conditions, form a phase portrait of the dynamical system.

The dynamical systems can be divided on conservative and dissipative systems. For the first type the whole energy of the system is conserved, for the second type can be energy losses. As concerns to our problem of research of massif state, which is in a state of oil recovery, the best model is such: heterogeneous, no stationary dissipative system. Nevertheless there can occur in the massif such local places, which will be described by a conservative dynamical model that is by a model of energetic equilibrium.

The analyses of the phase portrait of dynamical system allow us to make a conclusion about the system state during the period of observation. So, in conservative systems attracting sets do not exist. The set of phase space $\mathcal{R}_N$ is named attracting; to which trajectories tend with time, which begin in some its neighborhood. If in conservative system a periodical movement exists, thus such movements are infinitely many and they are defined by the initial value of energy. In dissipative systems the attracting sets can exist. Stationary oscillations for dissipative dynamical systems are not typical one. But in nonlinear systems a periodic asymptotic stable movement can exist, for which we have a mathematical image as a limited cycle, which is represented in the phase space as a closed line, to which all trajectories from some neighborhood of that line tend in time. We can conclude about the characteristic behavior of the system analyzing the form of phase portrait, by the way the "smooth" deformations of the phase space do not lead to quality changes of the system dynamics. That property is named as topologic equivalence of phase portraits. It allows analyzing the behavior of different dynamical systems from the unique point of view: on that base the set of dynamical systems can be divided on classes, inside of which the systems show an identical behavior. Mathematically "smooth deformation" of phase portrait is homeomorphous transformation of phase coordinates, for which new singular points can not occur, from the other hand – singular points can not vanish.

We had analyzed the seismological detailed information of space-time oscillations of state features of rock massif from the point of the theory of open dynamical systems [7 – 8]. We had revealed some synergetic features of the massif response on heavy man-made influences before a very intensive rock shock in the mine [9]. We defined a typical morphology of response phase trajectories of the massif, which is in the current time in stable state: on the phase plane we see a local area as a clew of twisted trajectories and small overshots from that clew with energies not more, than 10^5 joules. In some periods of time these overshots can be larger, than 10^6 joules up to 10^9 joules (see Figure1).

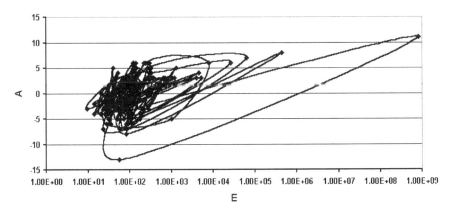

Figure 1. Phase portrait of the energies of massif responses during one of the most rock bursts on Tashtagol mine. Legend: E-energy in joules. A – d (LgE)/dt.

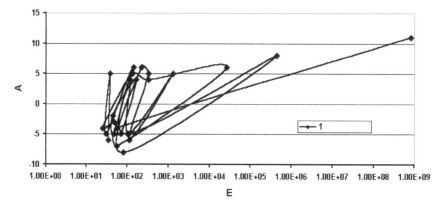

Figure 2. Phase portrait of the energies of massif responses before the one of the most rock bursts on Tashtagol mine. The legend of Figure 2. is identical to the legend of Figure 1.

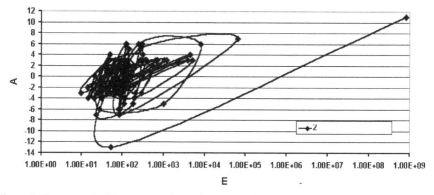

Figure 3. Phase portrait of the energies of massif responses after the same rock burst on Tashtagol mine. The legend of Figure 3. is identical to the legend of Figure 1.

Since the massif volume under investigation is the same and we research the process of it's activation and dissipation, we obviously see two mutual depending processes: the energy accumulation when phase point is near phase trajectories attracting area and resonant releasing of the accumulated energy. It is interesting to notice, that after the releasing the system returns to the same phase trajectory attracting area (see Figure 2 and 3).

In the book [8] is developed a new mathematical method for modeling of processes in local active continuum, which are energetically influenced from an outer energy source. The common causes of chaotization and stochastization of dynamical system movements are its losing of stability and exponential recession of near located phase trajectories together with its common boundedness and its common compression. The mathematical result coincides as a whole with the practical result (see Figure 2, 3): in the phase area the smaller attracting phase trajectories area exists where may occur an exponential recession of them (see Figure 2), then the movement character changes and the further movement of phase points lead to return to the same attracting area (see Figure 3.). These movements can occur in resonance or spontaneous cases.

The received results are of great significance because firstly we could find the coincidences between the mathematical theory of open systems and experimental results for natural objects with very complicated structure. On that base we developed a new processing method for the seismological information which can be used real-time for estimation of the disaster danger degree changing in mine massif.

The second feature of the state evolution is: the local massif volume does not immediately respond on the changing in environment stress state. Therefore it stores energy and then releases it with a high energy dynamical effect. It is very significant to define the time of reaction lagging, in spite of the influence on the massif can be assumed as elastic. The unique model which can explain that effect is a model of the massif with a hierarchic structure. We developed a mathematical algorithm using integral and integral-differential equations for 2D model for two problems in a frequency domain: diffraction of a sound wave and linear polarized transverse wave through an arbitrary hierarchy rank inclusion located in an N-layered medium. That algorithm differs from the fractal model approach by a freer selecting of heterogeneities position of each rank. And the second problem is solved in the dynamical approach. As higher amount of the hierarchic ranks as greater is the degree of nonlinearity of massif response and as longer can be the time of massif reaction lag of the influence [10].

In that paper integral equations and integral differential equations of 2D direct problem for the seismic field in the dynamical variant are derived and the joint analysis of the integral equations for 2D problems for electromagnetic and seismic fields had been provided. The received results can be used for definition of the complex criterions of achievement the research of high-complicated medium both with seismic and electromagnetic methods.

For the problem of sound diffraction on the 2D elastic heterogeneity, located in the j-th layer of the n-layered medium, using the approach from the papers [11, 12], we can derive the integral differential equation for the distribution of the potential for the vector of elastic displacements inside the heterogeneity. Using the second integral-differential presentation we can define the potential of the elastic displacements in the arbitrary layer, and then we can calculate the distribution of the vector of elastic displacements in the arbitrary layer. Let us compare the derived expressions with the solution of the diffraction problem for electromagnetic field in the frame of the same geometrical model. That case corresponds to the problem of exciting by a plane wave H – polarization, the solution of which is done in the paper [11]. Let us transform it to the form similarly to (1) and let us compare the derived equations for the solution of the inner 2D seismic and electromagnetic problem:

$$\frac{(k_{1ji}^2 - k_{1j}^2)}{2\pi} \iint\limits_{S_C} \varphi(M) G_{Sj}(M,M^0) d\tau_M + \frac{\sigma_{ja}}{\sigma_{ji}} \varphi^0(M^0) -$$

$$-\frac{(\sigma_{ja} - \sigma_{ji})}{\sigma_{ji} 2\pi} \oint\limits_C G_{Sj} \frac{\partial \varphi}{\partial n} dc = \varphi(M^0) \quad by\ M^0 \in S_C$$

$$\frac{\sigma_{ji}(k_{1ji}^2 - k_{1j}^2)}{\sigma(M^0) 2\pi} \iint\limits_{S_C} \varphi(M) G_{Sj}(M,M^0) d\tau_M + \varphi^0(M^0) -$$

$$-\frac{(\sigma_{ja} - \sigma_{ji})}{\sigma(M^0) 2\pi} \oint\limits_C G_{Sj} \frac{\partial \varphi}{\partial n} dc = \varphi(M^0) \quad by\ M^0 \notin S_C$$

(2)

$G_{Sj}(M,M^0)$ - the source function of seismic field for involved problem, $k_{1ji}^2 = \omega^2(\sigma_{ji} / \lambda_{ji})$; - index ji signs the membership to the features of the medium into the heterogeneity, λ – is a const of Lameux, σ -the density of the medium, ω-the round frequency, $\bar{u}_i = grad\varphi_i$; i=1,...j, ji,...n. $k^2(M^0) = i\omega\mu_0\sigma(M^0), \mu_0 = 4\pi10^{-7}\frac{2H}{M}, \sigma\ (M^0)$ - conductivity in the point M^0., i-the imaginary unit, $H_x(M^0)$ - the summarized component of magnetic field, $H_x^0(M^0)$ - the component of magnetic field in the layered medium without heterogeneity, $k_{ji}^2 = i\omega\mu_0\sigma_{ji}, k_i^2 = i\omega\mu_0\sigma_i$, σ_{ji} -conductivity into the heterogeneity, located into the j-the layer, σ_i - conductivity of the i-th layer of the n-layered medium, $G_m(M,M^0)$ - the Green function of the 2 – D problem for the case of H-polarization [18]. The difference in the boundary conditions for the seismic and electromagnetic problems lead to different types of equations: in the seismic case – to the integral-differential equation, in the electromagnetic caseв to the load integral equation of Fredholm of the second type.

$$\varphi(M^0) = \frac{(k_{1ji}^2 - k_{1j}^2)}{2\pi} \iint\limits_{S_C} \varphi(M) G_{Sp,j}(M,M^0) d\tau_M +$$

$$+\frac{(\sigma_{ji} - \sigma_{ja})}{\sigma_{ji} 2\pi} \oint\limits_C G_{Sp,j} \frac{\partial \varphi}{\partial n} dc + \frac{\sigma_{ja}}{\sigma_{ji}} \varphi^0(M^0)\ by\ M^0 \in S_C$$

$$H_x(M^0) = \frac{k_{ji}^2 - k_j^2}{2\pi} \iint\limits_{S_C} H_x(M) G_m(M,M^0) d\tau_M +$$

$$+\frac{k_{ji}^2 - k_j^2}{k_j^2 2\pi} \oint\limits_C H_x(M) \frac{\partial G_m}{\partial n} dc + \frac{k_{ji}^2}{k_j^2} H_x^0(M^0)\ by\ M^0 \in S_C$$

(3)

If for the solutions of the direct electromagnetic and seismic in dynamical variant problems we can establish the similarity in the explicit expressions for the components of electromagnetic and seismic fields by definite types of excitation then with complicating of the medium structure as can we see from the obtained result by the case of the seismic field linked with longitudinal waves the similarity vanishes. That means that the seismic information is additional to the electromagnetic information about the structure and state of the medium. For the problem of diffraction of a linearly polarized elastic transverse wave on the 2D heterogeneity located in the j-th layer of the n-layered medium, using the approach described in the paper [11] for the electromagnetic wave 2D problem (case H – polarization), (the geometric model is similar to a that described higher in the previous problem) we obtain the expressions as follows for the components of the displacement vector:

$$
\begin{aligned}
&\frac{(k_{2ji}^2 - k_{2j}^2)}{2\pi}\iint_{S_C} u_x(M)G_{Ss,j}(M,M^0)d\tau_M + \frac{\mu_{ja}}{\mu_{ji}}u_x^0(M^0) + \\
&+\frac{(\mu_{ja}-\mu_{ji})}{\mu_{ji}2\pi}\oint_C u_x(M)\frac{\partial G_{Ss,j}}{\partial n}dc = u_x(M^0) \;\; by\, M^0 \in S_C \\
&\frac{\mu_{ji}(k_{2ji}^2 - k_{2j}^2)}{\mu(M^0)2\pi}\iint_{S_C} u_x M)G_{Ss,j}(M,M^0)d\tau_M + u_x^0(M^0) + \\
&+\frac{(\mu_{ja}-\mu_{ji})}{\mu(M^0)2\pi}\oint_C u_x(M)\frac{\partial G_{Ss,j}}{\partial n}dc = u_x(M^0) \;\; by\, M^0 \notin S_C
\end{aligned}
\tag{4}
$$

The expressions (3) content the algorithm of seismic field simulation for distribution of transversal waves in the n-layered medium, which contain a 2D heterogeneity. The first expression is a Fredholm load integral equation of the second type the solution of which gives the distribution of the components of the elastic displacements vector inside the heterogeneity. The second of them is an integral expression for calculation of the elastic displacements vector in the arbitrary layer of the n-layered medium.

Comparing the expressions (3) with correspondingly for the electromagnetic field (H-polarization) (2) we see that there is a similarity of the integral structure of these expressions. The difference is only for the coefficients of corresponding terms in the expressions (2) and (3). That we can account by choosing the system of observation with one or another field. We must also account the difference of the medium response frequency dependence from seismic or electromagnetic excitation. But keeping within the similarity of the coefficients the seismic field, excited by transversal waves, and the electromagnetic field will contain the similar information about the structure of the heterogeneous medium and state, linked with it. Those results are confirmed by the natural experiments described in the papers [13 – 17]. Thus, it is showed that for more complicated, than horizontal-layered structures of the geological medium the similarity between the electromagnetic and seismic problems for longitudinal waves get broken. Therefore, these observations with two fields allow getting reciprocally additional information about the structure and especially about the state of the medium. These fields will differently reflect the peculiarities of the heterogeneous structures and response on the changing their state. If we can arrange seismic observations only with the transversal waves together with the magnetic component of electromagnetic one (H-polarization) in the 2D medium, it will be establish the similarity, which can be used by construction of mutual systems of observation for magneto-telluric soundings and deep seismic soundings on exchanged waves. For research of hierarchic medium we had developed an iterative algorithm for electromagnetic and seismic fields in the problem setting similar to analyzed higher for layered-block models with homogeneous inclusions [19].

3. Investigation of non-linear dynamics of rock massive, using seismological catalogue data and induction electromagnetic monitoring data in a rock burst mine

During the research of massif response on heavy explosions in some blocks of a rock burst hazard mine it had been derived some peculiarities of the rock massif behavior on different scale levels (Figures 4, 5). By exploitation of a concrete block the whole massif, mine field experiences the change of the stressed-deformed and phase state from explosion to explosion. The amounts of absorbed and dissipated by the massif energy are not equal to each other and therefore energy accumulation occurs inside the massif. The process of energy dissipating occurs with time delay and it strongly depends on the gradient of absorbing energy from mass explosions. Zones of dynamical calmness appear inside the massif. It is needed to trace such zones with use of seismological monitoring data and parameters described in [5]. After leaving out of the minimum of calmness it is needed during one or two weeks up to the moment of the technological crushing arrange the space-time active electromagnetic or seismic monitoring for revealing zones of potential non stability of the second rank. Such zones may appear after the mass crash explosion or after strong dynamical events.

These conclusions had been made using analysis of seismology data which is linked with the massif of concrete block mining. But the analysis of seismological data of the mine show that powerful dynamical events (rock bursts) can occur in more wide area than near of the block of mining and can be initiated in time delay. In the papers [5, 20] for the first time it had been analyzed the seismological detailed information from the synergetic position and the theory of open dynamical systems. Using the quality analysis of phase trajectories [21] the repeating regularities had been shown, consist of transitions in the massif state from chaotic to ordered and reverse.

For realization of that research data of the seismic catalogue of Tashtagol mine during two years from June 2006 to June 2008 had been used. As a data set we used the space-time coordinates for all dynamical events-responses of the massif occur in that time period inside the mine field and also explosions, which had been developed for massif outworking and values of energy which had been fixed by the seismological station. In our analysis we divided the whole mine field in two parts (figures 6, 7). The events-responses had been taken into account from horizons − 140 m, − 210 m, − 280 m, − 350 m. According to the catalogue, explosions had been provided on the south-east place – on horizons + 70 m, 0 m, − 70 m, on other places – on all listed higher horizons. The whole catalogue had been divided on two parts: north and south – for response events and for explosions occur in the south and north parts of the mine's field correspondingly. Between the explosions we summarized the precipitated energy of dynamical massif responses correspondently of the northern and southern parts.

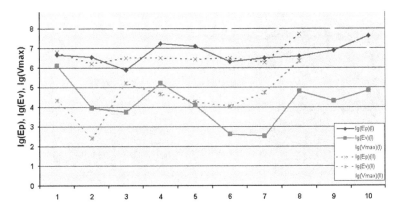

Ep – absorbed energy (joules), Ev-dissipated energy (joules), Vmax – the maximum dimension of the volume, where had been observed dynamical events.

Figure 4. The response of the massif by the man-made influence: 07. 10 – 23. 12 2001, block23, horizon(– 280/ – 210) (I) 19. 01 – 20. 03 2003, block24, horizon (– 280/ – 210)(II) X-axis– number of explosions

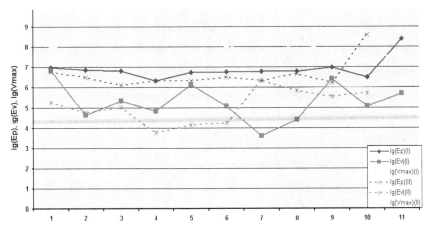

Figure 5. The response of the massif by the man-made influence: 27. 02 – 09. 07. 2000, block7, horizon (– 210/ – 240) (I), 21. 12 – 21. 03 2003 – 2004, block6, horizon (– 210/ – 140)(II). X-axis – number of explosions The legend of Figure 5 is the same as in Figure 4.

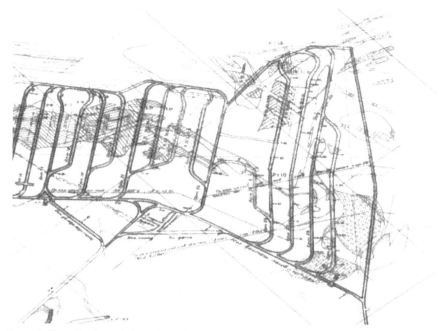

Figure 6. Plan of horizon – 210, southern place.

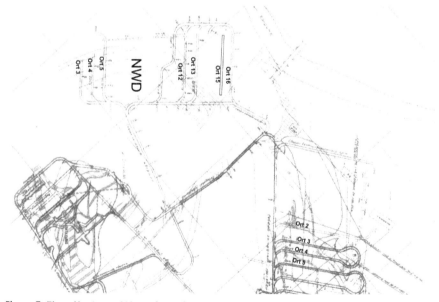

Figure 7. Plan of horizon – 210, northern place.

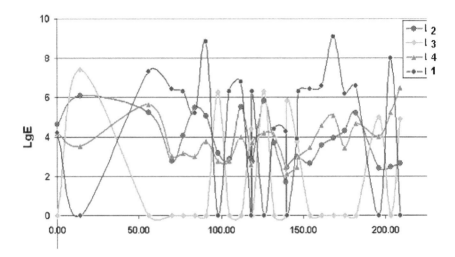

Figure 8. Distribution of absorbed (1 – 2) and dissipated (3 – 4) energy of the whole mine field during the period I 03. 06. 2006 – 13. 01. 2007. The horizontal axes is time per days

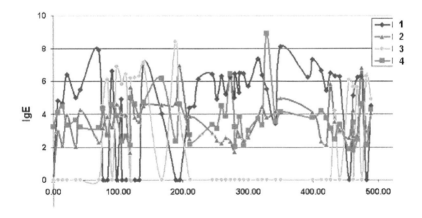

Figure 9. Distribution of absorbed (1 – 2) and dissipated (3 – 4) energy of the whole mine field during the period II 13. 01. 2007 – 17. 05. 2008. The horizontal axes is time per days.

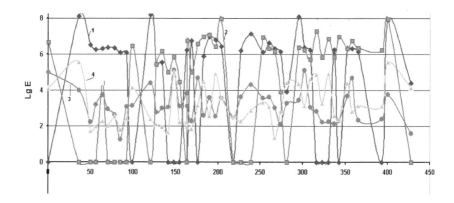

Figure 10. Distribution of absorbed (1 – 2) and dissipated (3 – 4) energy of the whole mine field during the period III 24. 05. 2008 – 26. 07. 2009. The horizontal axes is time per days

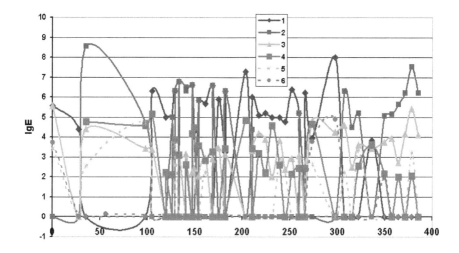

Figure 11. Distribution of absorbed (1 – 2) and extracted (3 – 6) energy of the whole mine field during the period IV 28. 06. 2009 – 18. 07. 2010. The horizontal axes is time per days

The whole interval of research had been divided into four periods: from 03. 06. 2006 – 13. 01. 2007 (period I), from 14. 01. 2007 – 17. 05. 2008 (period II), from 24. 05. 2008 – 26. 07. 2009 (period III), from 28. 06. 2009 – 18. 07. 2010 (period IV). (Figures 8 – 11).

We would like to note such peculiarities for the fourth period (Figure 11): during the period 100 days, beginning from 100 days from the beginning of the analyzed period and finishing by 200 days, the explosions had been provided in the northern and in the southern part of the mine field approximately of equal intensity, however the energy of massif response in the southern part is significantly larger, than from the northern part. During next 50 days the explosions had been provided in the southern part, but the energy of the massif response in the northern part and in the southern part are approximately equal. During the period from 300 days to 400 days the explosions had been provided mainly in the northern part of the mine field, the distribution of the massif response energy in the northern part practically corresponds to the distribution of the absorbed energy.

Period I						
Northern part(A)		Southern part (B)		Correlation coefficients		
Ep(joule)	Ev(joule)	Ep(joule)	Ev(joule)	R(Ep, Ev)(A)	R(Ep, Ev)(B)	R(Ev, Ev) (A, B)
3. $1 \cdot 10^7$	3. $1 \cdot 10^6$	2. $04 \cdot 10^9$	4. $11 \cdot 10^6$	0	0. 3	0. 34
Period II.						
2. $98 \cdot 10^8$	1. $67 \cdot 10^7$	3. $02 \cdot 10^8$	8. $2 \cdot 10^8$	– 0. 03	– 0. 12	– 0. 03
Period III						
2. $91 \cdot 10^8$	1. $3 \cdot 10^6$	5. $72 \cdot 10^8$	5. $81 \cdot 10^5$	0. 3	– 0. 005	0. 21

Table 1. Distribution of correlation coefficients during the I – III periods during the determined time intervals.

Days	Period 129 – 182	Period 203 – 259	Period 307 – 385
Northern part	R(Ea, Ed)=0. 02		R(Ea, Ed)=0. 52
Southern part	R(Ea, Ed)=0. 68	R(Ea, Ed)=0. 24	

Table 2. Distribution of correlation coefficients during the IV period during the determined time intervals.

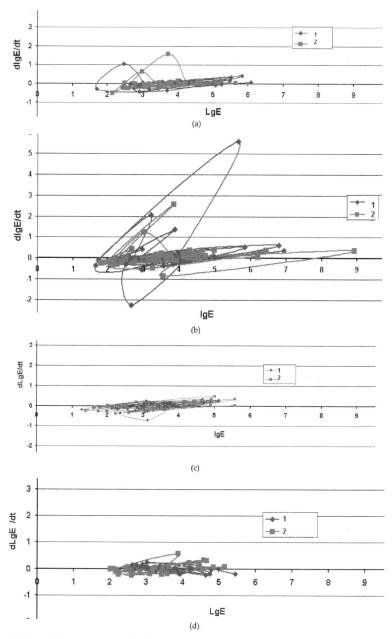

Figure 12. Phase diagrams: a) period 1, b) period 2, c) period 3, d) period 4., 1 – southern part, 2 – northern part.

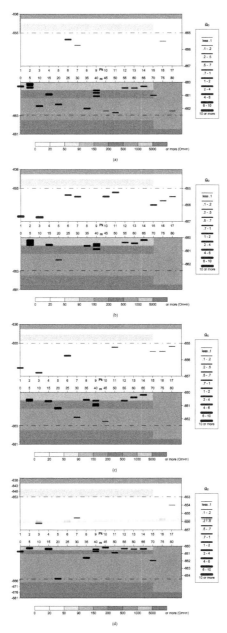

Figure 13. Geoelectrical sections along the profile horizon – 210, NWP, ort 3, frequency 10. 16 kHz. a) 9-th of July 2010, b) 16-th of July 2010, c) 23-th of July 2010, d) 26-th of July 2010, $\tilde{M}_0$ -intensity of discrete zones .

Here, looking the table 1, we can see the changing of nonlinearity in time of the massif state by its active blast effects. The practical massif state situation cannot be described by a linear model. Phase diagrams reflect the dependence of massif energy dissipation velocity from the irregularity in time active man-caused influence. So during the first period 03. 06. 2006 – 13. 01. 2007 the influence of that irregularity leads to dynamical events for the southern part 27. 08 2006. и 17. 09. 2006., in the northern part 10. 09. 2006, however the intensity of the dynamical events during that period did not exceed, than $5 \cdot 10^6$ joules. During the second period 14. 01. 2007 – 17. 05. 2008 the influence of the irregularity significantly increases compared to the first period, especially in the northern part of the mine and there occurred some dynamical events on 13. 05, 12. 08 and on 21. 10. 2007 in the southern part the most powerful rock burst occurred with energy amplitude – 10^9 joules.

The phase diagrams of the massif state during the two last periods show the identity of the massif state of the southern and northern parts during the last two years. The energy of responses and the velocity of its changing's has an identical character.

The successive cycle of induction active electromagnetic monitoring was provided in 2010 from 9-th of July to 26-th of July in the holes of the northern –west department and in some holes of the southern part of the mine's field. During the same time the man-caused works had been arranged in the block 4 – 5 of the northern part of the mine. The explosions had been achieved 04. 07 – Energy of the explosion 1. 7E+ 06 joules, 11. 07 – Energy of the explosion 3. 50E+ 07 joules, 18. 07 – Energy of the explosion 1. 7E+ 06 joules, 25. 07 – Energy of the explosion 1. 7E+ 06 joules, 01. 08 – Energy of the explosion 1. 4E+ 05 joules. The repeated electromagnetic observations had been achieved in the ort 3 NWD 9. 07, 16. 07, 23. 07 and 26. 07. (figure 10, (a-d)) The analyze of electromagnetic data during the 4 cycles of observation from 2007 to 2010 showed that the massif of the 3-d ort is more sensitive to the change of the stress-deformed state in the northern-west part of the mine field, which is caused by the influence outside it.

4. Geosynergetics approach for analyze of rock state – Theoretical and experimental results

The research of rock burst hazard massif of the mine Tashtagol was arranged using the approaches of the theory of open dynamical systems [6, 8, 20]. We would like to search the criterions of changing the regimes of dissipation for the real rock massif, which are under heavy man-caused influence. We used the data from the seismic catalogue from June 2006 to June 2008, used the space-time coordinates of whole dynamical events-responses of the massif, which occurred during that period and also explosions, which are fixed by the seismic station using the energy parameter [20]. The phase portraits of the massif state of the northern and southern places had been developed in the coordinates Ev (t) and d(Ev(t))/dt, t-time in the parts of days, Ev-seismic energy dissipated by the massif in joules.

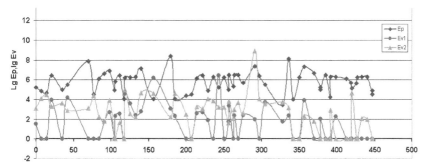

Figure 14. The distribution of absorbed (Ep) and dissipated energy (Ev) by the first and second part of the southern part of the massif during the second period of observation time 14. 01. 2007 – 17. 05. 2008. The horizontal ax is time per days.

Intervals	R(lgEp, lgEv2)		
1 – 21 0 – 179	– 0. 0062		
21 – 60 179 – 450	0. 240683		
23 – 42 200 – 342	0535269		
Intervals	R(lgEp, lgEv1)	R(lgEp, lgEv2)	R(lgEv2, lgEv1)
24 – 41 207 – 336	– 0. 00881		0. 12652
12 – 24 105 – 207	0. 30794	0. 13077	0. 70973
1 – 23 0 – 200			0. 26314

Table 3. Coefficients of correlation R for processes between absorption energy (Ep) and dissipation energy (Ev) in the massif for different time intervals (Ni), for the second period.

Here we divided the southern part of the mine's field on two parts: orts 13 – 16 – block (1) and 23 – 31 – block (2), horizons – 280, – 350 and used the data from the mines catalogue from 14-th January 2007 to 17-th of May 2008 (period 2).. From the results of the table 3 we see that the process of energy adsorption and dissipation in the researched blocks 1 and 2 as a rule is nonlinear, but the degree of nonlinearity changes in time. In the interval 23 – 42, which corresponds to the interval DT (200 – 342), the correlation coefficient between lg(Ep)(DT) and lg(Ev2)(DT) has the maximum value. That interval includes the process of preparation of the resonance release energy by the massif as a rock burst of the 9-th class. From the other side for the interval 12 – 42 we see the changing of the type of the correlation dependence between the functions R(lg(Ev2), lg(Ev1)): for the interval 12 – 24 the correlation coefficient corresponds to a linear correlation function, which reflects the elastic interaction between two blocks (orts 15 – 16) and (orts 27 – 29). By the way the relation between lg (Ep) and lg (Ev2) and lg(Ev1) is practically nonlinear. For the interval 24 – 41 the type of correlation function R(lg(Ep2), lg(Ep1)) changes nonlinear and the dependence between lg(Ep) and lg(Ev1) is practically absent (table 3). However before 48 days a rock burst of 6.4-th class occurs in the block 1 after the explosion of 5.2 classes in the block 2. Can we think that this rock burst was a foreshock for the rock shock of the 9-th class in the block 2? On that question we can answer only after providing detailed every day observations of electromagnetic active induction and deformation monitoring in the whole space of the 1-st and second blocks.

There is a deep principal difference between the mechanics of linear and nonlinear oscillations, which still persists by research weak nonlinear oscillations, which can be described by differential equations, which differ from linear equations only by presence of small terms, which begin to influence especially on the time intervals larger, than the period of oscillations. In the system there can be energy sources and absorbers, which produce and absorb very small work for one period of oscillations, but by prolonged actions their produced effect can be summarized and provide a sufficient influence on the oscillatory processing: decay, increase, stability. The small nonlinear terms can provide cumulative influence and damage the superposition principle, apart harmonics begin to influence on each other, and therefore the individual research of the behavior each harmonic oscillation apart cannot be done. The continuous waves can practically exist only in the case, if the system contains an energy source, which can compensate the energy decay, which appears as a result of existence of dissipative forces. Such source plays a role of negative friction. The oscillations which occur owing to a source, which influence has not constant period, are known as auto oscillations and any auto oscillatory system is described by a non linear differential equation.

Relaxation oscillations are widely distributed in the nature, for them oscillatory process has two stages: slow energy accumulation by the system and then energy relaxation, which take place almost immediately after the moment when potential threshold is over reached for that system.

5. Conclusions

One of the mathematical ideas about the common causes of chaotization and stochastization of dynamical system movements are its losses of stability and exponential recession of near located phase trajectories together with its common boundedness and its common compression [8]. The mathematical result coincides as a whole with the practical result: in the phase area the smaller attracting phase trajectories area exists where may occur an exponential recession of them, then the movement character changes and the further movement of phase points lead to return to the same attracting area. These movements may take place in resonance or spontaneous cases. The received practical results are significant because firstly we could find the coincidences with the mathematical theory between open systems and results of experiments in natural medium with very complicated structure. On this base we developed a new processing method for the seismological information which can be used real-time for estimation of the disaster danger degree changing in mine massif.

On the base of the constructed algorithm for calculation of the distribution of seismic waves into a medium with hierarchic inclusions located in an arbitrary layer of a horizontal layered elastic medium we can compute the stress components on each hierarchic level. This information we use to estimate the medium state analyzing the hierarchic structure and its changing. From the other side as higher the degree of hierarchic structure as larger becomes the degree of space nonlinearity of seismic field components distribution. This feature should be taken into account by interpretation to minimize linearization negative effects. From the received theoretical results we did a conclusion, that as higher is rank of hierarchy of the medium as less the similarity between seismic and electromagnetic results, and the obtained information has a independent sense, which underlines the complexity of the researched medium.

The analysis of experimental seismological and electromagnetic information showed the common additional information on different space-time scale levels of the state of rock massif, which is under energetic influence of mining explosions. It has revealed the change of nonlinearity in time of the massif state. The description of the massif behavior in the frame of the linear dynamical model does not correspond to the real practical situation. As it follows from the received results, the changing of the massif state: decreasing and increasing of its activization, does not depend on the space location of the explosion and on delay in time. It is needed to determine that delay function of its activization to be able to forecast the massif behavior. For that it is needed to continue obtaining and analyzing the complex information from passive and active seismic and electromagnetic detailed monitoring. For quantitative research of the behavior of different types of nonlinear dynamical systems we shall use the asymptotical methods for nonlinear mechanics, developed by N. M. Krilov, N. N. Bogolubov and Yu. A. Mitropolsky [bibliography in [23]]. However if we want to use the data from the seismic catalogue, we must convert energetic characteristics to forces and displacements. For that we need additional information about deformations, which occur in the massif by the explosions influence. The mathematical method, which had been developed by academician N. N. Bogolubov, allows us to get on with quantitative description of the causes of self excitation of the nonlinear mechanical system and the occurrence of the space-time local resonance in the system as a response to the outer influence. Using of common theoretical approaches [8, 23] and complex data: seismic catalogue data, deformation and induction electromagnetic monitoring data [20, 22] will allow us to formulate and solve the problem of forecasting the critical state of activated local place in rock massif.

That work is fulfilled by support of RFBR grant 10 – 05 – 00013a

Author details

O.A. Khachay*
Institute of Geophysics UB RAS, Yekaterinburg, Russia

A.Yu. Khachay and O.Yu. Khachay
Ural Federal University,
Institute of Mathematics and Computer Sciences, Yekaterinburg, Russia

6. References

[1] Physical mesomechanics and computer construction of materials. Novosibirsk: Nauka, SIFR; 1995. V. 1.

[2] Hachay O. A. The problem of the research of redistribution of stress and phase states of massif between high man-made influences. Mining information and analytic bulletin 2006; 5 109 – 115.

[3] Hachay O. A., Khachay O. Yu. Theoretical approaches for system of geophysical state control validation of geological medium by man-made influence. Mining information and analytic bulletin 2008; 1 161 – 169.

[4] Hachay O. A., Khachay O. Yu. Results of electromagnetic and seismic monitoring of the state of rock massif by use the approach of the open dynamical systems. Geophysical research abstracts 2009; 11 EGU2009 – 137.

[5] Hachay O. A. Synergetic events in geological medium and nonlinear features of wave propagation. Geophysical research abstracts 2009; 11 EGU2009 – 3684.

[6] Chulichkov A. I. Mathematical models of nonlinear dynamics. Moscow: Phismatlit; 2003.

[7] Malineckiy G. G. Mathematical base of synergetics. Moscow; LKI; 2007.

[8] Naimark Yu. I., Landa P. S. Stochastic and chaotic oscillations. Moscow; Knigniy dom "LIBROKOM"; 2009.

[9] Hachay O. A. Geosynergetics: theory, methods, experiments. In: Spichak V. V. (ed.) Electromagnetic research of the Earth. Moscow: Krasand; 2009. p. 138 – 153.

[10] Hachay O. A., Khachay A. Yu. Theory of integrating seismic and electromagnetic active methods for mapping and monitoring the state of 2D heterogeneities with hierarchic structure in a N-layered medium. In: Shkuratnik V. L., Fokin A. V., Mironov A. V., Didenkulov I. N. (eds) Physical acoustics, nonlinear acoustics, distribution and wave diffraction, geoacoustics : proceedings of the XX-th session of the Russian acoustic society 27 – 31 October 2008, Moscow. Russia. VI. Moscow: GEOS; 2008.

* Corresponding Author

[11] Dmitriev V. I. The diffraction of the plane electromagnetic field on cylindrical bodies, located in layered medium. In: Voevodin V. V. Methods of calculation and programming in layered medium. Moscow: MSU, 1965. V. 3. p. 307 – 316.

[12] Kupradze V. D. The boundary problems of the theory of waves and integral equations. Moscow-Leningrad: TTL. 1950,

[13] Khachay A. Yu. Algorithm of direct problem solution of dynamical seismic research by excitation with a vertical point force source, located in an arbitrary layer of n-layered isotropic elastic medium. In: Mazurov Vl. D., A. I. Smirnov (eds) Informatics and mathematical modeling. Yekaterinburg: USU; 2006. p. 279 – 310.

[14] Hachay O. A., Novgorodova E. N., Khachay A. Yu. Research of resolution of planshet electromagnetic method for active mapping and monitoring of heterogeneous geoelektrics medium. Fizika Zemli 2003; 1 30 – 41.

[15] Hachay O. A., Hinkina T. A., Bodin V. V. Preconditions of seismic and electromagnetic monitoring of nonstationary medium. Russian geophysical journal 2000; 17 – 18 83 – 89.

[16] Hachay O. A., Hinkina T. A., Bodin V. V. Research of the criterion of similarity for seismic and electromagnetic research in frequency-geometrical variant. In: Holshevnikov K. V., Shustov B. M. Zacharova P. E. (eds) Astronomical and geodetic research. Yekaterinburg: USU; 2001. p. 30 – 35.

[17] Hachay O. A., Druginin V. S., Karetin Yu. S., Bodin V. V., Novgorodova E. N., Khachay A. Yu., Zacharov I. B., Groznih M. V. The use of complex planshet seismic and electromagnetic method for solution of mapping subsurface heterogeneities problems. In: Solodilov L. N., Kostuchenko S. L., Yasuleyitch N. N. (eds.): Geophysics of the 21 – st century: proceedings of the third geophysical symposium, 22 – 24 February 2001,. Moscow, Russia. Moscow: Nauchniy Mir; 2001.

[18] Hachay. O. A. Mathematical modeling and interpretation of alternative electromagnetic field for heterogeneous crust and upper mantle of the Earth. Prof. thesis. Yekaterinburg: IGF UB RAS; 1994.

[19] Hachay O. A., Khachay O. Yu. Simulation of seismic and electromagnetic field in hierarchic heterogeneous structures. In: Martyshko P. S. (Ed): Geophysical research of the Ural and adjusted regions: proceedings of the international conference, 4 – 8 February 2008, Yekaterinburg, Russia. Yekaterinburg: OOO"IRA UTK", 2008.

[20] Hachay O. A., Khachay O. Yu., Klimko V. K., Shipeev O. V. Reflection of synergetic features of rock massif state under man-made influence in the data of seismological catalogue. Mining information and analytical bulletin 2010; 6 259 – 271.

[21] Hachay O. A., Khachay A. Yu. Construction of dynamical Model for Evolution of Rock Massive State as a Response on a Changing of stress-deformed State. Geophysical Research abstracts 2010; 12. EGU2010 – 2662.

[22] Hachay O. A., Khachay O. Yu, Klimko V. K., Shipeev O. Yu. The reflection of synergetic features in the response of geological medium on outer force actions. In: Jinghong F., Junqian Z., Haibo C., Zhaohui J.: Advances in heterogeneous Material Mechanics:

proceedings of the third international conference on heterogeneous material mechanics 22 – 26 May 2011, Shanghai, China. Lancaster, USA: DEStech Publications, Inc. 2011.

[23] Bogolubov N. N. Mathematics and nonlinear mechanics. Moscow: Nauka; 2005.

Chapter 6

Application of Multifractal and Joint Multifractal Analysis in Examining Soil Spatial Variation: A Review

Asim Biswas, Hamish P. Cresswell and Bing C. Si

Additional information is available at the end of the chapter

1. Introduction

Soil varies considerably from location to location [1] and the understanding of this variability has important applications in agriculture, environmental sciences, hydrology and earth sciences. For example, information about soil spatial variation is necessary for precision agriculture [2], environmental prediction [3], soil-landscape process modelling [4, 5], soil quality assessment [6, 7] and natural resources management. The quantification and characterization of soil spatial variability did not start until the latter half of the last century. Over the last four decades, systematic studies have identified the following characteristics of soil spatial variation.

- Spatial autocorrelation: Soil is a function of various environmental factors including climate, living organisms, relief, parent material, and time [8]. The individual or combined influence of these factors and various physical, chemical and biological processes produce different types of soil. Similar environmental factors and soil forming processes tend to have occurred at locations in close proximity to one another. It follows that soil properties measured at adjacent locations are likely to be more similar than properties measured at places located far apart [9, 10]. This is known as spatial autocorrelation or similarity.
- Scale dependence: The soil forming factors and processes can operate at different intensities and scales [4, 11]. Soil biological processes or the activity of micro- and macro-organisms can occur at very small scales affecting the formation of soils and the variability in soil properties. In contrast, atmospheric, geologic and climatic variability can determine the formation of soil and the variability of soil properties over a large area. The scale dependence of the environmental factors determines that soil spatial variability is also scale dependent.

- Periodicity: Sometimes the pattern of variation in soil attributes may repeat at different scales or over a certain distance beyond its spatial autocorrelation. The repetition or quasi-cyclic variation in soil can arise from the repeated features in topography, geology or parent material, tillage operation or cultivation practices [12-14]. This cyclic behaviour is known as the periodicity.

- Nonstationarity: The linear or nonlinear spatial trends in the controlling factors and processes may result in a change in soil properties that is gradual and predictable in space. This is referred to as the nonstationarity, which can arise from the effects of topography, lithology, parent material, climate and vegetation [15]. Non-stationarity can also occur where there are distinct strata in the variation, such as different types of soil. The mean and/or variance of soil properties in one stratum may differ from that of another.

- Nonlinearity: Often the effect from different factors and processes are non-additive in nature [16] and do not follow the principle of superposition. These are the characteristics of a nonlinear system [17], which can be explained by a simple example. Soil water storage is controlled by a number of factors [such as elevation, texture, vegetation, ...]. In nature, the overall response of soil water storage cannot be determined by simply observing the response of one factor at a time and subsequently adding the individual observational results together.

Earlier efforts in characterizing the soil spatial variability mainly focused on the spatial similarity in soil properties over a given area using soil classification based on soil survey and conventional statistics [18]. These methods assumed the variation to be random and spatially independent and did not quantify the variability of soil properties with respect to their spatial arrangement, spatial similarity or periodicity [2]. Geostatistical analysis, based on the theory of regionalised variables [19], has been used to characterize spatial similarity in soil properties [3, 20]. The spatial structure or the similarity information as a function of separation distance or scale [11] helps identify autocorrelation in replicating samples, reveals patterns in the data series and identifies the scale of major ongoing processes [21]. However, a necessary assumption in calculating a meaningful variogram, a cornerstone of geostatistical analysis, is that the variable is spatially stationary and the sum of squared differences depends only on the separation of measurements and not on their absolute locations [2, 22].

Different processes controlling the variability in soil properties can contribute differently. Variance contribution from individual processes towards the total variance can be evaluated by transforming the spatial domain information to the frequency domain. The processes that operate at a very fine scale have high spatial frequencies, whereas processes that operate at very broad scales have low spatial frequencies [23]. The contribution of different processes at different scales and their repeated behaviour can be quantified using spectral analysis [12, 24]. Spectral analysis approximates a spatial series by a sum of sine and cosine functions. Each of the functions has an amplitude and a frequency or period. The squared amplitude at a given frequency is equal to the variance contribution of the frequency component to the total variance in the spatial series [25-27]. It deals with the global information or the mean state and cannot examine localized variations or long-term trends [23]. Moreover, spectral analysis assumes that a spatial series is second-order stationary (i.e. the mean and variance of the series are finite and constant). This assumption is generally stricter than the intrinsic stationarity assumption of geostatistics.

Wavelet transformation [28] has been used to examine both long term trends and localized features in soil spatial variation. It enables analysis of multi-scale stationary and/or nonstationary soil spatial variation over a finite spatial domain and has become popular for examining scale and location dependent soil spatial variation [23, 29-33]. A review of the application of the wavelet transform in soil science can be found in Biswas and Si [34]. This method has been extremely useful in examining nonstationary soil spatial variation. However, in examining nonlinear soil spatial variation, the appropriateness of the wavelet transform has been questioned. The Hilbert-Huang transform was introduced into soil science to examine the nonstationary and nonlinear soil spatial variation together [32, 35]. These methods have been useful in examining soil spatial variability. However, they only deal with how the second moment of a variable changes with scales or frequencies.

For normally distributed variables, the second moment plus the average provide a complete description of the variability in the spatial series. The inherent soil variability and the extrinsic factors can cause orders of magnitude of variation in measured soil property values. The presence of intermittent high and low data values often result in distributions deviated from the normal. For these type of distributions (e.g., left skewed distribution), higher moments are needed for a complete description of the variability in the measured property. If we define the qth moment of Z as $\langle z^q \rangle$, then when q is positive, the qth moments magnify the effect of larger numbers in the spatial series Z and diminish the effect of smaller numbers in the data series. On the other hand, when q is negative, the qth moments magnify the effect of small numbers and diminish the effect of large numbers in the spatial series Z. In this way, by varying the order of the moments, we can look at the magnitude of data values and visualize a better picture of the data series. There is a need to summarize how these moments change with scales or the scaling properties of these moments to compare and simulate spatially-variable soil properties.

Soil properties sometimes vary in space in an irregular manner or exhibit no deterministic patterns. If the irregularity in a variable's distribution remains statistically similar at all studied scales, the variable is assumed to be self-similar [36]. Self similarity is closely associated with the transfer of information from one scale to another (scaling). Exploring self-similarity or inherent differences in the scaling properties is important in order to understand the nature of the spatial variability in soil properties. Fractal theory originated by Mandelbrot [37] can be used to investigate and quantitatively characterize spatial variability over a range of spatial scales. Mandelbrot's early work in the area of geophysics, specifically the characterization of coast lines, showed that the patterns observed at different scales could be related to each other by a power function, whose exponent was called as the fractal dimension. The fractal geometry offered both descriptive and predictive opportunities in the field of soil science [38]. It has been useful in providing a unique quantitative framework for integrating soil biological, chemical and physical phenomena over a range of spatial and temporal scales.

Most of the fractal theory applications in soil science have used a single fractal dimension to characterize the spatial variability over a range of scales [4, 38]. This is known as a monofractal approach - it assumes that the soil spatial distribution can be uniquely characterized by a single fractal dimension. However, the spatial variability in soil properties is the result of the individual or combined influence of soil physical, chemical and biological processes operating at different intensities and scales [11]. Therefore, monofractal distributions are not likely to be prevalent in the landscape [39]. A single fractal dimension will not always be sufficient to represent complex and heterogeneous behavior of soil spatial variations. An extension of the monofractal approach was introduced in soil science to describe data with a set of fractal dimensions instead of a single value. A spectrum can be prepared by combining all the fractal dimensions and is known as the multifractal spectrum. The method of characterizing variability based on the multifractal spectrum is known as multifractal analysis [40]. Multifractal behavior is associated with a system where the underlying physics are governed by a random multiplicative process (i.e., successive division of a measure and its spatial support based on a given rule). Therefore, multifractal behavior implies that a statistically self-similar measure can be represented as a combination of interwoven fractal dimensions with corresponding scaling exponents. The multifractal parameters are generally independent of the size of the studied objects [41] and do not assume any specific distribution in the data [42]. Multifractal analysis can transform irregular data into a compact form and amplify small differences among the variables [43, 44]. It uses a wider range of statistical moments, providing a much deeper insight into the data variability structure compared to the methods that use only first two statistical moments. Multifractal analysis can therefore be used to characterize the variability and heterogeneity in soil properties over a range of spatial scales [44-47]. While the multifractal analysis characterizes the spatial variability in a variable, joint multifractal analysis characterizes the joint distribution of two variables along a common spatial support. It can provide information on the relationships between two variables across different spatial scales [47, 48]. The objective of this chapter is to demonstrate what multifractal and joint multifractal analysis can do in dealing with spatial data series from the published soil science literature. In the next section we briefly describe multifractal and joint multifractal analysis methodology including the working steps and then proceed to describe some applications of these methods in soil science. Finally we close the chapter with a discussion on future prospects for multifractal and joint multifractal methods in soil science within a brief Conclusions Section.

2. Methods

2.1. Multifractal analysis

Multifractal analysis can be used to characterize the scaling property of a soil variable measured in a direction (such as a transect) as the mass (or value) distribution of a statistical measure on a geometric or spatial support. The geometric or spatial support indicates the extent of the sampling. This can be achieved by dividing the length of the transect into smaller and smaller segments based on a rule that generates a self similar segmentation. One such rule is the binomial multiplicative cascade [36] that can divide a unit interval associated with a unit mass, M (a normalized probability distribution of a variable or a measured distribution as used in a generalized case) into two segments of equal length. This also partitions the associated mass into two fractions, $h \times M$ and $(1-h) \times M$ and assign them as the left and right segments, respectively. The parameter h is a random variable values between 0 and 1. The new subsets and its associated mass are successively divided into two parts following the same rule. The differences among the subsets are identified using a wide range of statistical moments, which can then be used to determine the multifractal spectra of the measures [36, 48]. Therefore, the multifractal analysis focuses on how the measure of mass varies with box size or scale and provides physical insights at all scales without any ad hoc parameterization or homogeneity assumptions [49]. A detailed description of the multifractal theory can be found in reference [36, 50, and 51]. In this section, for brevity, we will only summarize the computational techniques and concepts commonly used in examining the soil spatial variation.

Past research has indicated that certain properties of a spatial series decrease with increase in scale, following a power law relationship. For example, when all or part of the variogram follows a power law equation of the form $\gamma(h) \approx h^{-\varsigma}$ the data are scaling in that range—i.e., there is certain degree of statistical scale-invariance [52].

However, the semivariogram only measures the scaling properties of the second moment. Similarly, we can do the same thing for the higher moments such as third, fourth, and so on. Will the scaling properties be the same at higher moments or change with the order of the moments, q? If the scaling properties do not change with q, then we say the spatial series is monofractal, i.e., it only requires a single scaling coefficient to transfer information from one scale to another. If the scaling coefficient changes with q, then the spatial series is multifractal, i.e., it requires multiple scaling coefficients for transferring information over scales. For a spatial series, the scale-invariant mass exponent (structure function), $\tau(q)$, is defined as [46]:

$$\langle [\Delta Z(x)]^q \rangle \propto x^{\tau(q)} \tag{1}$$

where the symbol "$\propto$" indicates proportionality, Z is the spatial series, and x is the lag distance. If the plot of $\tau(q)$ vs. q [or $\tau(q)$ curve; Fig. 1a] has a single slope (i.e., a straight line), then the series is a simple scaling (monofractal) type. If the $\tau(q)$ curve is nonlinear and convex (facing downward), then the series is a multiscaling (multifractal) type (Fig. 1b).

The type of scaling can be examined from the degree of fractality (i.e., monofractal or multifractal) by comparing the $\tau(q)$ curve with a reference curve or a theoretical model (Fig. 1a). One such reference curve (similar to the monofractal type of scaling) or the theoretical model was proposed by Schertzer and Lovejoy [49], which is known as the universal multifractal model or UM model. The UM model simulates an empirical moment scaling function of a cascade process assuming the conservation of mean value. The UM model can be used to compare and characterize the observed scaling properties as reference to the monofractal behavior or scaling. The similarity/dissimilarity can be examined from the goodness-of-fit between the $\tau(q)$ curve and the UM model using the chi-square test. Statistically significant difference between the curves indicates multifractal scaling and non-significant difference indicates monofractal scaling. The degree of monofractal/multifractal can also be examined by calculating the deviation of the $\tau(q)$ curve from the UM model [47, 53]. Large sum of squared difference between the curves indicates multifractal behavior and small sum of squared difference indicates monofractal behavior. The slopes of the regression line fitted to the $\tau(q)$ curve (also referred as single fit) can be compared to the slopes of the UM model. Significant difference in the slope indicates multifractal behavior. The $\tau(q)$ can even be fitted to two regression lines; one for $q>0$ and another $q<0$ (also referred as segmented fit). The difference between the means of slopes from segmented fits (for positive and negative q values) can be tested using the Student's t test. Significant difference between the slopes indicates nonlinearity in the $\tau(q)$ curve and thus multifractal behavior [47, 53].

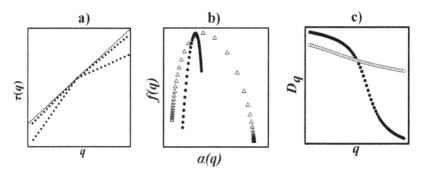

Figure 1. A typical of a) mass exponent graph (or $\tau(q)$ curve), b) multifractal spectrum (or $f(q)$ curve) and c) generalized dimension graph (or Dq curve) used to identify the monofractal and multifractal behavior in any property.

Research has indicated that the qth order normalized probability measures of a variable (also known as the partition function), $\mu(q,\varepsilon)$, vary with the scale size, ε in a manner similar to Eq. (1) [36, 51], i.e.,

$$\mu_i(q,\varepsilon) = \frac{\left[p_i(\varepsilon)\right]^q}{\sum_i \left[p_i(\varepsilon)\right]^q} \propto (\varepsilon / L)^{\tau(q)} \tag{2}$$

where $p_i(\varepsilon)$ is the probability of a measure in the ith segment of size ε units, calculated by dividing the value of the variable in the segment by the whole support length of L units. In simpler terms, $p_i(\varepsilon)$ measures the concentration of a variable of interest (e.g., clay content, organic carbon, etc.) in a given segment relative to the whole support length. The $\tau(q)$ function in Eq. (2) is given a new name, 'the mass exponent,' because it relates the probability of mass distribution in a given segment to the size of the segment (scale) and is used widely in multifractal analysis.

The multifractal spectrum, $f(q)$ (Fig. 1b), which is the fractal dimension of the subsets of segments of size ε units with a coarse Hölder exponent (local scaling indices) of α if calculated to the limit as $\varepsilon \rightarrow 0$, can be expressed as [36, 50],

$$f(q) = \lim_{\varepsilon \rightarrow 0} \left(\log\left(\frac{\varepsilon}{L}\right) \right)^{-1} \sum_i \mu_i(q,\varepsilon) \log \mu_i(q,\varepsilon) \tag{3}$$

and the local scaling indices, α, are given by

$$\alpha(q) = \lim_{\varepsilon \rightarrow 0} \left(\log\left(\frac{\varepsilon}{L}\right) \right)^{-1} \sum_i \mu_i(q,\varepsilon) \log p_i(\varepsilon) \tag{4}$$

Note that $f(\alpha)$ can also be determined through the Legendre transform of the $\tau(q)$ curve:

$$f(\alpha(q)) = q\alpha(q) - \tau(q) \tag{5}$$

The multifractal spectrum is a powerful tool in portraying the variability in the scaling properties of the measures (e.g., clay content, organic carbon, etc.). The spectrum also enables us to examine the local scaling property of soil variable. The width of the spectrum ($\alpha_{max} - \alpha_{min}$) is used to examine the heterogeneity in the local scaling indices (Fig. 1b). The wider the spectrum, higher the heterogeneity in the distribution of the soil variable. Similarly, the height of the spectrum corresponds to the dimension of the scaling indices. Small $f(q)$ values correspond to rare events (extreme values in the distribution), whereas the largest value is the capacity dimension that is obtained at $q = 0$.

For many practical applications indicator parameters are selected and used, in addition to the multifractal spectrum ($f(q)$ vs. $\alpha(q)$), to describe the scaling property and variability of a process. Generalized dimensions, D_q, (Fig. 1c) are often used to provide indicator parameters. The D_q of a multifractal measure is calculated as

$$D_q = \frac{1}{q-1} \lim_{\varepsilon \rightarrow 0} \frac{\log \sum_i p_i(\varepsilon)}{\log(\varepsilon)} \tag{6}$$

The D_q value at $q = 0$, D_0, is called the capacity dimension or the box counting dimension of the geometric support of the measure. The value at $q = 1$, D_1, is referred to as the information dimension and provides information about the degree of heterogeneity in the distribution of the measure – this is analogous to the entropy of an open system in thermodynamics [54]. Sometimes, D_1 is also known as the entropy dimension. A value of D_1 close to unity indicates the evenness of measures over the sets of a given cell size, whereas a value close to 0 indicates a subset of scale in which the irregularities are concentrated. The D_2, known as the correlation dimension, is associated with the correlation function and measures the average distribution density of the measure [55]. For a distribution, with simple scaling (monofractal), the D_1 and D_2 become similar to the D_0, the capacity dimension. The value of $D_0 = D_1 = D_2$ indicates that the distribution exhibit perfect self-similarity and is homogeneous in nature (Fig. 1c). However, for multifractal type scaling the order generally becomes $D_0 > D_1 > D_2$. The D_1/D_0 value can also be used to describe the heterogeneity in the distribution [56]. A D_1/D_0 value equal to 1 indicates exact monoscaling of the distribution, which means that all fractions take equal values at different scales.

2.1.1. Multifractal analysis steps

1. Calculate the probability measure (p) from a linear distance for a transect, or from a rectangle for an area. Regarding the minimum number of samples required to carry on the analysis, the multifractal analysis method has the flexibility over other methods such as Fourier transform or wavelet transform, which generally require a larger dataset with regular interval between samples.

   a. For each n = 2, 4, 8, 16, . . . until the unit is not dividable. To calculate the probability measure (p), sum up the values of all the points in the segment, and divide by the total of all segments $\left(p_i(\varepsilon) = \dfrac{m(\varepsilon, i)}{M} \right)$ where i is the index for i^{th} segment when the whole transect is divided into n segments with a unit length = $\varepsilon (i = 1, \cdots, n)$; $m(\varepsilon, i)$ is the sum of measurements at all points in the ith segment of length ε; M is the total of all points along the transect.

   b. For certain q values or statistical moments (e.g., -20 to 20, which are selected based on the nature of the data to be analyzed), calculate $\mu_i(q, \varepsilon)$ using Eq. (2).

      a. Calculate $Ff(q, \varepsilon) = \sum_i \mu_i(q, \varepsilon) \log \mu_i(q, \varepsilon)$

      b. Calculate $F\alpha(q, \varepsilon) = \sum_i \mu_i(q, \varepsilon) \log p_i(\varepsilon)$

2. Check if the partition function $\mu(q, \varepsilon)$ obeys the power law or if the log-log plot of the partition function and distance is linear. If they are not, then they are not multifractal and no further multifractal analysis is needed. If they are, continue the following.

3. Calculate $\tau(q)$ as the intercept of linear regression of $\log\left[\mu_i(q, \varepsilon) \right]$ vs. $\log(\varepsilon)$.

4. Calculate $f(q)$ as the intercept of linear regression of $Ff(q, \varepsilon)$ vs. $\log(\varepsilon)$.

5. Calculate $\alpha(q)$ as the intercept of linear regression of $F\alpha(q, \varepsilon)$ vs. $\log(\varepsilon)$.

6. Calculate D_q as the intercept of the linear regression of $FD(q, \varepsilon)$ vs. $\log(\varepsilon)$.

7. Plot $\tau(q)$ as a function q, $f(q)$ as a function of $\alpha(q)$, and D_q as a function of q.

All of these calculations can be implemented in MathCad, MATLAB, or SAS.

2.2. Joint multifractal analysis

While the multifractal analysis characterizes the distribution of a single variable along its spatial support, the joint multifractal analysis can be used to characterize the joint distribution of two or more variables along a common spatial support. Similar to the multifractal analysis, the length of the datasets (for example, 2 datasets) is divided into a number of segments of size ε. The probability of the measure of the ith segment of the first variable is $P_i(\varepsilon) \propto (\varepsilon/L)^{\alpha}$ and for the second variable is $R_i(\varepsilon) \propto (\varepsilon/L)^{\beta}$, where α and β are the local singularity strength corresponding to $P_i(\varepsilon)$ and $R_i(\varepsilon)$, respectively. The partition function (the normalized μ measures) for the joint distribution of $P_i(\varepsilon)$ and $R_i(\varepsilon)$, weighted by the real numbers q and t can be calculated as [47, 50, 51];

$$\mu_i(q,t,\varepsilon) = \frac{p_i(\varepsilon)^q \cdot r_i(\varepsilon)^t}{\sum\limits_{j=1}^{N(\varepsilon)} \left[p_j(\varepsilon)^q \cdot r_j(\varepsilon)^t \right]} \tag{7}$$

The local scaling indices (coarse Hölder exponents) with respect to two probability measures $P_i(\varepsilon)$ and $R_i(\varepsilon)$, which are represented by $\alpha(q,t)$ and $\beta(q,t)$ respectively, are calculated as follows:

$$\alpha(q,t) = -\left[\log(N(\varepsilon)) \right]^{-1} \sum_{i=1}^{N(\varepsilon)} \left[\mu_i(q,t,\varepsilon) \cdot \log(p_i(\varepsilon)) \right] \tag{8}$$

$$\beta(q,t) = -\left[\log(N(\varepsilon)) \right]^{-1} \sum_{i=1}^{N(\varepsilon)} \left[\mu_i(q,t,\varepsilon) \cdot \log(r_i(\varepsilon)) \right] \tag{9}$$

The dimension (i.e., $f(\alpha,\beta)$) of the set on which $\alpha(q,t)$ and $\beta(q,t)$ are the mean local exponents of both measures is given by

$$f(\alpha,\beta) = -\left[\log(N(\varepsilon)) \right]^{-1} \sum_{i=1}^{N(\varepsilon)} \left[\mu_i(q,t,\varepsilon) \cdot \log(\mu_i(q,t,\varepsilon)) \right] \tag{10}$$

When q or t is set to zero, the joint partition function shown in Eq. [7] reduces to the partition function of a single variable, and hence the joint multifractal spectrum defined by Eq. [10] becomes the spectrum of a single variable. When both q and t are set to zero, the maximum $f(\alpha,\beta)$ is attained, which is the dimension if all the segments contain the same concentration of mass. Different pairs of α and β can be scanned by varying the parameters q and t. Because, high q or t values magnify large values in the data and negative q or t values magnify small values in the data, by varying q or t, we can examine the distribution of high or low values (different intensity levels) of one variable with respect to that of the other variable. Pearson correlation analysis can be used to quantitatively illustrate the variation of the scaling exponents of one variable with respect to another variable across similar moment orders. Because $f(\alpha,\beta)$ represents the frequency of the occurrence of a certain value of α and a certain value of β, high values of $f(\alpha,\beta)$ signifies a strong association between the values of α and the values of β. By perturbing q and t, we can examine the association of similar values (highs vs. highs or lows vs. lows) of α and β as well as dissimilar values (highs vs. lows) of α and β. Generally, a contour graph is used to represent the joint dimensions, $f(\alpha,\beta)$ of the pair of variables (Fig. 2). The bottom left part of the contour graph shows the joint dimension of the high data values of the two variables, while the top right part represents the low data values [57]. The diagonal contour with low stretch indicates strong correlation between values corresponding to the variables in the vertical and horizontal axis (Fig. 2).

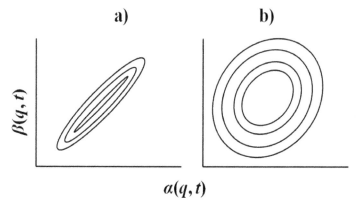

Figure 2. A typical of multifractal spectra for joint distribution of two variables with a) strong correlation and b) weak correlation.

2.2.1. Joint multifractal analysis steps

1. Calculate the probability measure (p) for variable Y from a linear distance for a transect or a rectangle from an area.

    a. For each n = 2, 4, 8, 16, . . . until the unit is not dividable, calculate the probability measure (p), sum up values of all the points in the segment, and divide by the total of all segments $\left(p_i(\varepsilon) = \dfrac{m_Y(\varepsilon,i)}{M_Y} \right)$ where i is the index for i^{th} segment when the whole transect is divided into n segments with a unit length $= \varepsilon(i=1,\cdots,n)$;

    Similarly, we can calculate the probability measure for the variable Z: $r_i(\varepsilon) = \dfrac{m_Z(i)}{M_Z}$.

    b. For certain q values (e.g., -20 to 20) and t values (e.g., -20 to 20), calculate $\mu_i(q, t, \varepsilon)$ using Eq. (7).

    c. Calculate $Ff(q,t,\varepsilon) = \sum_i \mu_i(q,t,\varepsilon)\log \mu_i(q,t,\varepsilon)$

d.   Calculate $F\alpha(q,t,\varepsilon) = \sum_i \mu_i(q,t,\varepsilon)\log p_i(\varepsilon)$

e.   Calculate $F\beta(q,t,\varepsilon) = \sum_i \mu_i(q,t,\varepsilon)\log r_i(\varepsilon)$

2.   Calculate $\tau(q,\, t)$ as the intercept of linear regression of $\log\left[\mu(q,t,\varepsilon)\right]$ vs. $\log(\varepsilon)$ (Eq. (7)).

3.   Calculate $f(q,\, t)$ as the intercept of linear regression of $Ff(q,t,\varepsilon)$ vs. $\log(\varepsilon)$ (Eq. (8)).

4.   Calculate $\alpha(q,\, t)$ as the intercept of linear regression of $F\alpha(q,t,\varepsilon)$ vs. $\log(\varepsilon)$ (Eq. (9)).

5.   Calculate $\beta(q,\, t)$ as the intercept of linear regression of $F\beta(q,t,\varepsilon)$ vs. $\log(\varepsilon)$ (Eq. (10)).

6.   Plot $\tau(q)$ as a function q and contour plot of $f(\alpha,\, \beta)$.

Again, all these calculations can be implemented in MathCad, MATLAB, or SAS.

2.3. Comments on multifractal and joint multifractal analysis

- Multifractal analysis for two-dimensional or three dimensional fields is the same as for a transect. However, the support ε becomes an area or volume.
- Multifractal analysis, like spectral analysis, is based on the global statistical properties of spatial series. Therefore, localized information is lost, which is different from wavelet analysis. However, multifractal analysis provides information regarding the higher moments and how the higher moments change with scale.
- Multifractal analysis does not require regularly spaced samples. Any sampling scheme can be analyzed by multifractal analysis.
- For highly spatially variable fields, the probability p for some locations may be very small or even zero. As such, the negative power of p can be very large and the partition function will be dominated by this single value. In this case, the multifractal method may diverge and the process of division needs to be stopped.
- Joint multifractal analysis can be used for the simultaneous analysis of several multifractal measures existing on the same geometric support, and hence for quantifying the relationships between the measurements studied. Joint multifractal analysis is based on the assumption that the individual variable is multifractal.
- Simulation of a synthetic field according to the measured multifractal distribution may enhance our understanding of the effects, spatial scaling, and spatial variability of soil properties on various soil processes.

3. Application of multifractal and joint multifractal analysis in soil science

Fractal theory [37] has been used to investigate and quantitatively characterize spatial variability over a large range of measurement scales in different fields of geosciences including soil science [4, 58]. A detailed review of the applications of fractal theory in soil science can be found in reference [38]. The fractal theory applications in soil science used monofractal approach, which assumes that the soil spatial distribution can be uniquely characterized by a single fractal dimension. However, a single fractal dimension might not be sufficient to represent complex and heterogeneous behavior of soil spatial variations. Multifractal analysis, which uses a set of fractal dimensions instead of one, is useful in characterizing complex soil spatial variability. Use of multifractal analysis in explaining soil spatial variability started during the late 1990's but only become popular in soil science after 2000. Among the earlier works, Folorunso et al. [59] used multifractal theory for analyzing spatial distribution of soil surface strength and Muller [60] used multifractal analysis for characterizing pore space in chalk materials. Since then, multifractal analysis has been used in soil science to study various issues including spatial variability of soil properties (e.g., physical, chemical, hydraulic), soil groups and pedotaxa, soil particle size distribution, soil surface roughness and microtopography, effect of tillage activities, crop yields, soil porosity and pore size distribution, flow and transport of water and chemical in soil, infiltration, and downscaling soil water information from satellite images. While multifractal analysis is used to characterize the spatial variability of soil properties over a range of scales, the joint multifractal analysis can be used in soil science to characterize the joint distributions between soil properties over a range of scales. In this section we review previous applications of multifractal and joint multifractal analysis in soil science.

3.1. Multifractal analysis

Various soil properties have shown multifractal behavior and the variability in those soil properties has been characterized using multifractal analysis. For example, Kravchenko et al. [44] was one of the first to report the multifractal nature of soil-test phosphorus (P), potassium (K), exchangeable calcium (Ca), magnesium (Mg) and cation exchange capacity (CEC). The study used a 259 ha agricultural field in central Illinois, USA. One set of grid size was used and the authors reported only one unique set of multifractal spectra for each soil property. Multifractal parameters were studied over a range of moment orders (q) from -15 to 15. The minimum value of multifractal parameters α (scaling indices) and $f(\alpha)$ (multifractal spectra), i.e., α_{min} and $f(\alpha_{min})$, corresponded to $q = 15$, and the maximum values, α_{max} and $f(\alpha_{max})$ corresponded to $q = -15$. A high value of $\alpha_{max} - \alpha_{min}$ indicated a wider opening of the multifractal spectrum and thus the multifractal nature of the soil properties except organic matter (OM) content and soil pH (Fig. 3) [44]. A very small value of $\alpha_{max} - \alpha_{min}$ indicated the monofractal nature of OM and pH (Fig. 3). The generalized fractal dimensions were also calculated for all positive q values. An excellent fit between the theoretical fractal dimension model and the $D(q)$ curve also indicated the multifractal nature of the soil properties except OM and soil pH. A high correlation between the multifractal parameters and exploratory statistics (e.g., coefficient of variations, skewness, kurtosis) or geostatistical parameters (e.g., nugget, range) provided comprehensive information on the major aspects of data variability [44]. Multifractal spectra of the soil properties studied carried a large amount of spatial information and allowed quantitative differentiation between the soil variability patterns. Kravchenko et al. [44] highlighted the usefulness of multifractal parameters of soil properties in the interpolation and mapping of those properties. Interpolation methods based on the multifractal scaling relationship improves the mapping of soil properties.

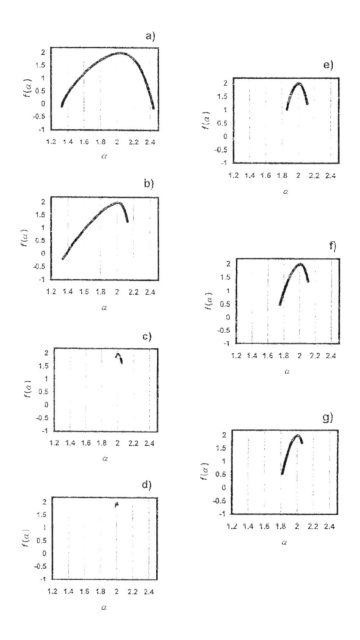

Figure 3. Multifractal spectra, $f(q)$, for a) phosphorus, b)potassium, c) organic matter, d) pH, e) exchangeable Ca, f) exchangeable Mg, and g) cation exchange capacity (adopted from 44).

Various other studies from different parts of the world also indicate the multifractal behavior of soil properties. Caniego et al. [61] studied spatial variability in soil electrical conductivity (EC), pH, OM content and depth of Bt horizon along three transects - two short transects (each 33 m long) with high intensity sampling (sampling interval 25 cm), and a long transect (3 km) with 40 m sampling interval. Though the authors reported the multifractal nature of the soil properties along the long transect, the variability along the short transects were more homogeneous. The variability along the short transect was explained by simple random fractal noise (close to a white noise) and thus a single scaling index was considered sufficient to transform information from one scale to another [61]. The variability in the long transect might have a deterministic component reflecting changing geological features and differences in soil forming factors with distance. The presence of nonlinearity together with the environmental features along the long transect resulted in the observed multifractal behavior of soil properties. Monofractal behavior of sand and silt was reported by Wang et al. [62]. Zeleke and Si [52, 57] characterized the distribution of bulk density (Db), sand content (SA) and to some extent silt (SI) content as monofractal along a gently sloping 384 m long transect in semi-arid central Saskatchewan, Canada. However, these authors did report multifractal behavior of clay (CL), saturated hydraulic conductivity (Ks) and organic carbon (OC) from the same study. For example, Fig. 4 shows the $\tau(q)$ (mass exponents) and $f(q)$ (multifractal spectra) curve of the soil variables studied by Zeleke and Si [52]. The linearity of the $\tau(q)$ curve of Db, SA and SI indicated a monofractal nature, while the convex shape of the $\tau(q)$ curve of CL, OC, and Ks indicated multifractal behavior. Similarly, a large value for $\alpha_{max} - \alpha_{min}$ of CL, OC, and Ks, and thus the wide opening of the curve, indicated the multifractal nature of these properties (Fig. 4) [52]. Wang et al. [62] also reported highly multifractal behavior of OC and CL content along a transect and monofractal behavior of SA and SI. A decrease in the variability of EC was reported with an increase in the EC value itself [63].

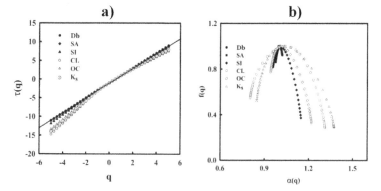

Figure 4. a) Mass exponents of different variables. The solid line is the UM model passing through $\tau(0)$. b) The multifractal spectra of different variables. The q = -6 to 6 at 0.5 increments at both cases. Db is the bulk density, SA is the sand content, SI is the silt content, CL is the clay content, OC is the organic carbon content, Ks is the saturated hydraulic conductivity. (adopted from 52)

The multifractal behavior of effective saturated hydraulic conductivity in a two-dimensional spatial field was reported by Koirala et al. [64] from a numerical simulation study. The authors used a normalized mass fraction of a randomized multifractal Sierpinski carpet to represent the areal distribution of saturated hydraulic conductivities in an aquifer. The effective saturated hydraulic conductivity was related to the generalized dimensions of the multifractal field. This relationship may be helpful to predict the effective saturated hydraulic conductivity of an aquifer from an empirical Dq spectra determined by the multifractal analysis [64]. The multifractal nature of the hydraulic conductivity was also reported by Liu and Molz [46].

The multifractal behavior of soil water content [53, 63, 65, 66], soil water retention at different suction [57, 62, 67, 68] and theoretical water retention parameters (e.g., van Genuchten α and n) [62, 67-71] was also reported in literature. The volumetric as well as gravimetric soil water content was found to exhibit multifractal behavior [65]. The variability in soil water content decreases with the increase in soil water and increases with the increase in the representative area of the measurement [66]. Though the water retention at saturation showed monofractal behavior, water retention at -30 kPa and -1500 kPa showed multifractal behavior [57, 67]. Filling up the pore with water during saturation irrespective of the size of the pores might result in characteristic that is persistent over a range of scales. However, at -30 kPa and -1500 kPa, the relation between pore size and pore water retention might result in multifractal behavior [57, 67]. While the van Genuchten [72] water retention parameter n (considered as relating to pore size distribution) showed monofractal behavior, the parameter α (considered as the inverse of air entry potential) showed weak multifractal behavior [62, 69]. Similar behavior of water retention parameters was reported by Liu et al. [70, 71]. However, Guo et al. [68] reported multifractal behavior of both water retention parameters, n and α.

The fractal behavior of soil water estimates derived from the satellite images were reported earlier [73, 74]. The relationship between the variance in soil water and the area of measurement or aggregation area (or scale) indicated self-similarity or scale invariance in soil water estimates. Later this relationship was used to develop models for downscaling information on soil water estimates from satellite images. A number of studies have developed and used multifractal models to downscale soil water estimates from satellite images to represent small areas [75-79]. Often the passive remote sensing images provide estimates of soil water for a large area (coarse spatial resolution; e.g. 25 50 km). Downscaling models are necessary tools to characterize and reproduce soil water heterogeneity from the remotely sensed estimates. The presence of multifractal behavior helped developing downscaling models [76].

Prediction of soil water storage as affected by soil microtopography or microrelief can also be made using the multifractal approach [80-81]. Soil surface roughness and microtopography showed multifractal behavior, the quantification of which help characterizing spatial features in topography and water storage in micro-depressions. The soil surface roughness created by tillage operations can determine soil strength, penetration of roots and susceptibility to wind and water erosion. The multifractal behavior of soil surface roughness can be used to explain the structural complexity created by tillage operations and the effect of various tillage implements. Various studies have reported the multifractal behavior of soil surface roughness [80-85]. Generally, these studies measured the soil surface roughness in a very small area. The degree of multifractality was found to increase at higher statistical moments [84]. The variability in the distribution of soil units also showed multifractal behavior [86-87]. Information on the variability within and between soil pedological units (pedotaxa) shows promise in analyzing and characterizing the complexity of soil development at multiple scales.

A large number of studies reported the multifractal behavior of soil particle size distribution [56, 88- 95]. Most of the studies used soil particle size distributions measured using laser diffraction. One of the studies also used a piece-wise fractal model for very fine and coarse particle sizes [95]. Multifractal behavior of soil particles sizes at different land can be used to characterize soil physical health and its quality [90].

Distribution of soil particles size determines the porosity, which in turn determines the flow and transport of water and chemicals in soil. Soil thin sections have been used to study the pore arrangements and distributions in soil [96-98]. The pore geometry showed multifractal behavior, which is useful for classification of soil structure and determining the fluid flow parameters [96]. The multifractal nature of soil porosity was used to identify the representative elementary area using photographs of soil and confocal microscopy [99]. The use of two dimensional binary or grayscale images [96-98] of soil thin sections helped in characterizing the pore structure and movement of water in soil. With the advancement of technology, the images from Magnetic Resonance Imaging (MRI) [102] or Computer Tomography (CT) [101, 102] helped in characterizing the soil as a porous media and the fluid flow through it. Three-dimensional images of soil systems helped to identify soil pores and their connectivity in three-dimensions, which in turn helped understand the movement of water in soil and the characterization of preferential flow paths [100]. The multifractal behavior of soil pore systems is often used to develop models for creating simulated media, which helps to study flow and transport of water and chemicals in soil or other porous media [103].

The infiltration of water into soil often shows multifractal behavior. Use of dye is a common approach to study the preferential flow paths of water in soil. The distribution of dye along with the infiltration water helped illustrate multifractal behaviour [104-106]. The multifractal nature of infiltration helped in identifying the input parameters for various rainfall-runoff models [107-109]. Perfect et al. [109] used a multifractal model to develop a simulated media for upscaling effective saturated hydraulic conductivity.

The radioactive cesium (^{137}Cs) fallout at small spatial scales has been found to be multifractal in nature [111]. Grubish [111] reviewed research on the ^{137}Cs fallout and reported that the multifractal nature stemmed from the erosion and deposition of soil materials, which showed a log-normal distribution. The behavior of soil chemical processes such as nitrogen absorption isotherms was also found to be multifractal [110]. The authors studied the isotherm characteristics from 19 soil profiles in a tropical climate. The asymmetric singularity spectra (shifted to the left) indicated the highly heterogeneous and anisotropic distribution of the measure (Fig. 5). The generalized dimension spectra (Dq) was used to discriminate among soil groups with contrasting properties, arising from different pedological processes such as Ferralsols (Latosols) and soil with argillic (Bt) horizons and/or an abrupt textural change.

Various studies also reported the multifractal behavior of soil landscape properties. For example, Wang et al. [112] found the multifractal nature of the spatial and temporal patterns of land uses in China. The multifractal pattern was also found in different terrain indices. While the upslope area and wetness index were multifractal, the relative elevation was monofractal [47, 113]. This may be due to inclusion of multi-scale characteristics in calculating secondary terrain indices such as the wetness index. The spatial variability of crop yield also found to exhibit multifractal in nature [47, 113-115]. A stochastic simulation study showed that the spatial variability in the production of wheat crop was multifractal, while the production of corn was monofractal [115].

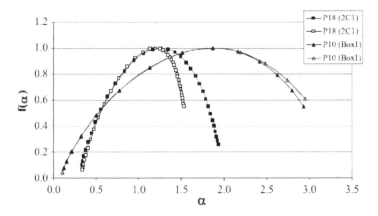

Figure 5. A multifractal spectrum generated using duplicate samples from two soil horizons that exhibit the lowest and the highest mean Hölder exponents of order zero (adopted from 110).

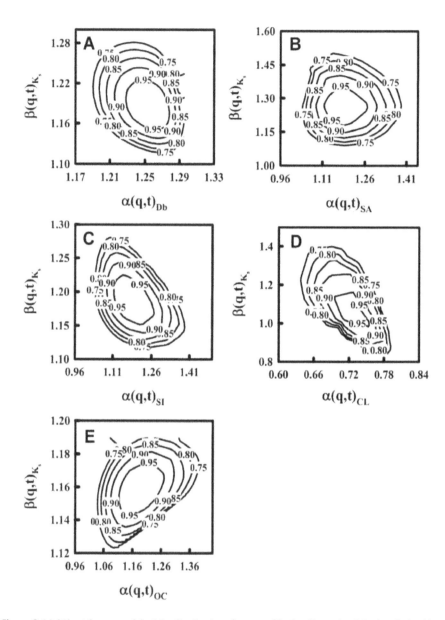

Figure 6. Multifractal spectra of the joint distribution of saturated hydraulic conductivity (vertical axis) and other soil properties (horizontal). Contour lines show the joint scaling dimensions of the variables. Db is the bulk density, SA is the sand content, SI is the silt content, CL is the clay content, OC is the organic carbon content, Ks is the saturated hydraulic conductivity (adopted from 52)

3.2. Application of joint multifractal analysis in soil science

Joint multifractal analysis has been used to characterize the joint distribution between two soil properties over a range of scales. Often soil hydraulic properties are predicted from more easily measured soil physical properties using pedotransfer functions. The scale dependence of soil physical properties often creates issues in the prediction. A large body of literature used joint multifractal analysis to characterize the joint distribution between soil physical and hydraulic properties [52, 57, 62, 63, 66, 68-71]. Zeleke and Si [52] studied the relationship between the saturated hydraulic conductivity and various soil physical properties. Figure 5 shows the contour lines for the joint scaling dimensions of K_s and Db. There appear to be some relationships between the scaling dimensions of K_s and Db for both high and low data values, which is evident from the slightly diagonal feature of the plots and the high correlation coefficients between the scaling indices of the two variables (Fig. 6). The highest correlation coefficient ($r = -0.57$) between scaling indices of Db and K_s was obtained for the high data values of the two variables. However, for joint multifractal spectra of SA and K_s, the contour line is nearly concentric, indicating there are no preferred associations between values of exponents for K_s and sand content. Therefore in this case, prediction of the scaling properties of K_s from those of sand content would not be recommended [52].

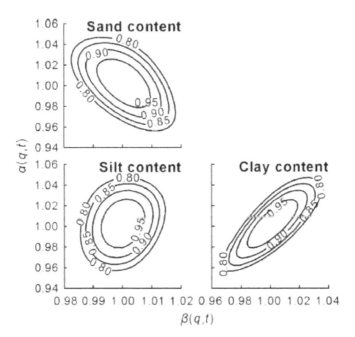

Figure 7. Multifractal spectra of the joint distribution of van Genuchten α (vertical axis) and the soil texture (horizental axis). (adopted from 69).

Various theoretical hydraulic parameters (e.g., van Genuchten α and n) are used to characterize soil water retention capacity. Often these parameters are predicted from soil physical properties potentially introducing issues of scale dependency. The joint distribution between theoretical soil hydraulic parameters and various soil physical properties were studied using joint multifractal analysis [69]. The contour plot for van Genuchten α and sand content are elliptical and tilted to the left (Fig. 7). However, the small difference between the long axis and short axis of the ellipse indicated that the van Genuchten α was not strongly correlated to the sand content over multiple scales. Similarly, there was weak correlation between the van Genuchten α and the silt. However, the elliptical contour tilted to the right indicated a strong positive correlation between the van Genuchten α and clay content (Fig. 7). The small diameter of the contour across the tilt compared with that along the tilt indicated low variability in the joint scaling exponents between the van Genuchten α and clay content. This means that the van Genuchten α and clay content had strong relationship at all intensity levels (high vs. high and low vs. low data values). This may be because the dominant factor of aggregation in loam soil is clay content. Characterizing the relationship at multiple scales using joint multifractal analysis is different from the traditional correlation analysis, which only explains the variability at the scale of measurement. However, variability analyses at a single scale may not accurately reflect the spatial patterns and processes of the studied variables.

The relationship between crop yield and various terrain indices over a range of scales were also studied using joint multifractal analysis [47, 48]. Quantification of the spatial variability in the crop yield and the yield affecting factors has important implications in precision agriculture. Topography is often considered as one of the most important factors affecting yields, and topographical data are much easier to obtain than time and labor-consuming measurements of soil properties. Kravchenko et al. [48] used the joint multifractal analysis to differentiate between yield-distributions corresponding to field locations with high and low slopes. The authors observed larger yields at low slope locations and a wide range of yields at sites with moderate and high slopes during four growing seasons with moderate and dry weather conditions. During the wet growing season, lower yields prevailed at locations with low slopes [48]. The applicability of joint multifractal analysis to study crop yield and topography relationships and characterize spatially distributed data has also been reported by Zeleke and Si [47]. The authors reported that the distribution of yield and biomass was better reflected in upslope length than other topographic indices such as wetness index and relative elevation. This implies that the upslope length can be used as a guideline for varying production inputs such as fertilizer, especially when detailed recommendations at a given scale of interest are not available. The strong relationship between the upslope length and the crop yield implies that one can make any site specific prediction of the final yield (before the harvest) using the upslope length regardless of scale [47]. The variability in crop yield with respect to the nitrate nitrogen level in soil was also studied using joint multifractal analysis [114]. Various authors also developed models to better predict the crop yield based on the multifractal nature of the crop yield and terrain indices [115].

Leaf area index (LAI) is often used as the growth indicator for plants. Banerjee et al. [113] studied the relationship between LAI of pasture and various terrain indices including relative elevation, wetness index, upslope length and distance. There was high correlation between LAI and the wetness index and upslope length over a range of scales. The study indicated that the spatial heterogeneity in LAI is well reflected in the variability of wetness index and upslope length [113] enabling the prediction of variability in LAI and thus the grass production from the variability in terrain indices.

4. Conclusion and future prospects

In this chapter, we reviewed various applications of multifractal and joint multifractal analysis to spatial studies in soil science. Spatial variation of soil has spatial dependence, periodicity, nonstationarity, nonlinearity and many other characteristics. Geostatistics quantifies spatial dependence and spectral analysis can analyze the scale information, but loses spatial information. The wavelet transform can be used to examine the spatial variability of nonstationary series, and the Hilbert-Huang transform can examine the spatial variability of nonstationary and nonlinear series concurrently. These methods can only deal with the second moment of a variable change with scales or frequencies. For a normally distributed variable, the second moment and the average provide a complete description of the variability in a spatial series. However a second order moment can only provide a poor characterization of the variability that occurs in non-normal distributions. For distributions other than normal, we need higher moments for complete characterization of soil spatial variability.

Fractal theory can be used to investigate and quantitatively characterize spatial variability over a large range of measurement scales. According to the fractal theory, the properties that are observed at different scales are related to each other (self-similar) by a power function, whose exponent is called the fractal dimension. When a single fractal dimension is used to characterize soil spatial variability, it is considered as monofractal. However, soil variability can occur in different intensities at different scales and a single fractal dimension may not be sufficient. The multifractal theory implies that a statistically self-similar measure can be represented as a combination of interwoven fractal dimensions with corresponding scaling exponents. A multifractal spectrum is prepared by combining all the fractal dimensions, which can be used to characterize the variability and heterogeneity in a variable over a range of scales. A large number of fractal sets provides in-depth characterization of soil spatial variability compared to that using a single fractal dimension. The multifractal approach is independent of the size of the studied variable and does not require any assumption about the data following any specific distribution. Therefore, the multifractal approach has been used to characterize spatial variability of soil properties, soil groups, soil water, crop yield, and terrain properties. The multifractal nature of the variability has been used to downscale soil water estimates from satellite images. The observed self-similar nature of pore sizes and their distribution has been used to characterize the flow and transport of water and chemicals through soil.

While multifractal theory is used to characterize the variability in any one soil property, the joint multifractal theory was used to characterize the joint distribution of two variables along a common spatial support. Various soil properties are difficult to measure as well as time, cost and labor intensive, while other properties are easy to measure. The variability in one property that can be easily measured can be used to predict the variability in other properties that are difficult to measure. The joint distribution of soil hydraulic properties (e.g., saturated hydraulic conductivity, theoretical hydraulic parameters) with soil physical properties (sand, silt, clay, bulk density, organic carbon) has been used to predict the variability in soil hydraulic properties. Similarly, crop yield has been predicted from its relationship with various terrain indices and other chemical properties.

Multifractal and joint multifractal analyses are versatile tools for characterizing soil spatial variability at multiple scales. Use of the multifractal methods with other approaches such as wavelet analysis would provide extensive information on soil spatial variability at different scales and locations. Multifractal analysis deals with global information and the wavelet transform can deal with local information. Joining the methods would help complete characterization of soil spatial variability [67, 116-118]. Often the localized trends and nonstationarities in the spatial series create a challenge in the scaling analysis. Moreover, nonstationarity can introduce a superficial scaling in the soil properties. In this situation, the intrinsic scaling property from variation in a given soil attribute needs to be separated from the undue influence of larger scale trends. Use of detrended fluctuation analysis together with multifractal analysis provides an opportunity to characterize the intrinsic variability of soil properties at field scale resulting from the interaction of all underlying processes [53]. Similarly, the use of a multifractal detrended moving-average can be an opportunity to identify the intrinsic scales of variability [119], and multifractal detrended moving-average cross-correlation analysis can be used to analyze and compare the variability between two soil attributes [120].

Author details

Asim Biswas* and Hamish P. Cresswell
CSIRO Land and Water, Canberra, ACT, Australia

Bing C. Si
Department of Soil Science, University of Saskatchewan, Saskatchewan, Canada

Acknowledgement

The funding for this work was provided through a CSIRO Land and Water post doctoral fellowship. The thoughtful comments from Dr. Gallant are highly appreciated.

5. References

[1] Nielsen DR, Biggar JW, Erh KT (1973) Spatial variation of field measured soil water properties. Hilgardia 42: 214–259.

---

* Corresponding Author

[2] Goderya FS (1998) Field scale variations in soil properties for spatially variable control: a review. J. soil contam. 7: 243–264.

[3] Trangmar BB, Yost RS, Uehara G (1985) Application of geostatistics to spatial studies of soil properties. Adv. agro. 38: 45–94.

[4] Burrough PA (1983) Multiscale sources of spatial variation in soil. 1. The application of fractal concepts to nested levels of soil variation. J. soil sci. 34: 577–597.

[5] Corwin DL, Hopmans J, de Rooij GH (2006) From field- to landscape-scale vadose zone processes: scale issues, modeling, and monitoring. Vadose zone j. 5: 129–139.

[6] Heuvelink GBM, Pebesma EJ (1999) Spatial aggregation and soil process modeling. Geoderma 89: 47–65.

[7] McBratney AB, Bishop TFA, Teliatnikov IS (2000) Two soil profile reconstruction techniques. Geoderma 97: 209–221.

[8] Jenny H (1941) Factors of Soil Formation. McGraw-Hill, New York, NY.

[9] Gajem YM, Warrick AW, Myers DE (1981) Spatial dependence of physical properties of a Typic Torrifluvent soil. Soil sci. soc. am. j. 45: 709–715.

[10] Oliver MA (1987) Geostatistics and its application to soil science. Soil use manag. 3: 8–20.

[11] Goovaerts P (1998) Geostatistical tools for characterizing the spatial variability of microbiological and physico-chemical soil properties. Biol. fert. soils 27: 315–334.

[12] Kachanoski RG, Rolston DE, de Jong E (1985) Spatial and spectral relationships of soil properties and microtopography. 1. Density and thickness of A-horizon. Soil sci. soc. am. j. 49: 804–812.

[13] Van Wesenbeeck IJ, Kachanoski RG, Rolston DE (1988) Temporal persistence of spatial patterns of soil-water content in the tilled layer under a corn crop. Soil sci. soc. am. j. 52: 934–941.

[14] Perfect E, Caron J (2002) Spectral analysis of tillage-induced differences in soil spatial variation. Soil sci. soc. am. j. 66: 1587–1595.

[15] Van Wambeke A, Dulal R (1978) Diversity of soils in the Tropics. Am. soc. agron. spec. publ. 34: 13–28.

[16] Yan R, Gao RX (2007) A tour of the Hilbert-Huang transform: An empirical tool for signal analysis. IEEE instru. meas. mag. 1094-6969/07: 40–45.

[17] Pai PF, Palazotto AN (2008) Detection and identification of nonlinearities by amplitude and frequency modulation analysis. Mech. syst. signal process. 22: 1107–1132.

[18] Beckett PHT, Webster R (1971) Soil variability: a review. Soils fert. 34: 1–15.

[19] Matheron G (1963) Principles of geostatistics. Econ. geol. 51: 1246–1266.

[20] Burgess TM, Webster R (1980) Optimal interpolation and isarithmic mapping of soil properties. I. The semivariogram and punctual kriging. J. soil sci. 31: 315–331.

[21] Si BC, Kachanoski RG, Reynolds RD (2007) Analysis of soil variation. In: Gregorich EG, editor, Soil sampling and methods of analysis. CRC Press, New York, NY. pp. 1163–1191.

[22] Oliver MA, Webster R (1991) How geostatistics can help you. Soil use manag. 27: 206–217.

[23] Si BC (2003) Spatial and scale dependent soil hydraulic properties: a wavelet approach. In Pachepsky Y, Radcliffe DW, Magdi Selim H, editors, Scaling method in soil physics. CRC Press, New York, NY. pp. 163–178.

[24] Koopmans LH (1974) The spectral analysis of time series. Academic Press. New York. USA. pp. 366.

[25] Webster R (1977) Spectral analysis of gilgai soil. Aust. j. soil res. 15: 191–204.

[26] Shumway RH, Stoffer DS (2000) Time series analysis and its applications. Springer, New York, NY.

[27] Brillinger DR (2001) Time series: Data analysis and theory. Soc. Ind. Appl. Math. Philadelphia, PA. pp. 540.

[28] Mallat S (1999) A wavelet tour of signal processing. Second edition. Academic Press, New York, NY.

[29] Lark RM, Webster R (1999) Analysis and elucidation of soil variation using wavelets. Eur. j. soil sci. 50: 185–206.

[30] Lark RM, Webster R (2001) Changes in variance and correlation of soil properties with scale and location: analysis using an adapted maximal overlap discrete wavelet transform. Eur. j. soil sci. 52: 547–562.

[31] Si BC, Farrell RE (2004) Scale dependent relationships between wheat yield and topographic indices: A wavelet approach. Soil sci. soc. am. j. 68: 577–588.

[32] Biswas A, Si BC (2011). Identifying scale specific controls of soil water storage in a Hummocky landscape using wavelet coherency. Geoderma 165, 50–59.

[33] Biswas A, Tallon LK, Si BC (2009) Scale-specific relationships between soil properties: Hilbert-Huang transform. Pedometron 28: 17–20.

[34] Biswas A, Si BC (2011) Application of continuous wavelet transform in examining soil spatial variation: a review. Math. geosci. 43: 379–396.

[35] Biswas A, Si BC (2011) Revealing the controls of soil water storage at different scales in a Hummocky landscape. Soil sci. soc. am. j. 75: 1295–1306.

[36] Evertsz CJG, Mandelbrot BB (1992) Multifractal measures (Appendix B). In: Peitgen HO, Jurgens H, Saupe D, editors, Chaos and fractals. Springler-Verlag, New York. pp. 922-953.

[37] Mandelbrot BB (1982) The fractal geometry of nature. W.H. Freeman, San Francisco, CA.

[38] Perfect E, Kay BD (1995) Applications of fractals in soil and tillage research: A review. Soil till. res. 36: 1–20.

[39] Milne BT (1991) Lessons from applying fractal models to landscape patterns. In: Turner MG, Gardner RH, editors, Quantitative methods in landscape ecology. Springer-Verlag, Berlin. pp. 200–235.

[40] Frisch U, Parisi G (1985) On the singularity structure of fully developed turbulence. In: Gil M, editor, Turbulence and predictability in geophysical fluid dynamics. North Holland Publ. Co., Amsterdam. pp. 84–88.

[41] Cox LB, Wang JSY (1993) Fractal surfaces: Measurement and applications in earth sciences. Fractals 1: 87–117.

[42] Scheuring I, Riedi RH (1994) Application of multifractals to the analysis of vegetation pattern. J. veg. sci. 5: 489–496.

[43] Folorunso OA, Puente CE, Rolston DE, Pinzon JE (1994) Statistical and fractal evaluation of the spatial characteristics of soil surface strength. Soil sci. soc. am. j. 58: 284–294.

[44] Kravchenko AN, Boast CW, Bullock DG (1999) Multifractal analysis of soil spatial variability. Agro. j. 91: 1033–1041.

[45] Olsson J, Niemczynowicz J, Berndtsson R (1993) Fractal analysis of high-resolution rainfall time series. J. geophys. res. 98: 23265–23274.

[46] Liu HH, Molz FJ (1997) Multifractal analyses of hydraulic conductivity distributions. Water resour. res. 33: 2483–2488.

[47] Zeleke TB, Si BC (2004) Scaling properties of topographic indices and crop yield: Multifractal and joint multifractal approaches. Agro. j. 96: 1082–1090.

[48] Kravchenko AN, Bullock DG, Boast CW (2000) Joint multifractal analysis of crop yield and terrain slope. Agro. j. 92: 1279–1290.

[49] Schertzer D, Lovejoy S (1997) Universal multifractals do exist! : Comments. J. app. meteor. 36: 1296–1303.

[50] Chhabra AB, Meneveau C, Jensen RV, Sreenivasan KR (1989) Direct determination of the f(α) singularity spectrum and its application to fully developed turbulence. Phys. rev. A 40: 5284–5294.

[51] Meneveau C, Sreenivasan KR, Kailasnath P, Fan MS (1990) Joint multifractal analyses: Theory and application to turbulence. Phys. rev. A 41: 894–913.

[52] Zeleke TB, Si BC (2005) Scaling relationships between saturated hydraulic conductivity and soil physical properties. Soil sci. soc. am. j. 69: 1691–1702.

[53] Biswas A, Zeleke TB, Si BC (2012) Multifractal detrended fluctuation analysis in examining scaling properties of the spatial patterns of soil water storage. Nonlin. proc. geophy. 19: 227–238.

[54] Voss RF (1988). In. Peitgen HO, Saupe D, editors, Fractals in nature: from characterization to simulation, in the science of fractal images, Springer, New York, pp. 21–70.

[55] Grassberger P, Procaccia I (1983) Measuring the strangeness of strange attractors. Physica D 9: 189–208.

[56] Montero E (2005) Renyi dimensions analysis of soil particle-size distributions. Ecol. model. 182: 305–315.

[57] Zeleke TB, Si BC (2006) Characterizing scale-dependent spatial relationships between soil properties using multifractal techniques. Geoderma 134: 440–452.

[58] Eghball B, Schepers JS, Negahban M, Schlemmer MR (2003) Spatial and temporal variability of soil nitrate and corn yield: Multifractal analysis. Agro. j. 95: 339–346.

[59] Folorunso OA, Puente CE, Rolston DE, Pinzon JE (1994) Statistical and fractal evaluation of the spatial characteristics of soil surface strength. Soil sci. soc. am. j. 58: 284–294.

[60] Muller J (1996) Characterization of pore space in chalk by multifractal analysis. J. hydrol. 187: 215–222.

[61] Caniego FJ, Espejo R, Martin MA, San Jose F (2005) Multifractal scaling of soil spatial variability. Ecol. model. 182: 291–303.

[62] Wang Z, Shu Q, Liu Z, Si B (2009) Scaling analysis of soil water retention parameters and physical properties of a Chinese agricultural soil. Aust. j. soil res. 47: 821–827.

[63] Liu J, Ma X, Zhang Z (2010a) Multifractal study on spatial variability of soil water and salt and its scale effect. Trans. chinese soc. agril. eng. 26: 81–86.

[64] Koirala SR, Perfect E, Gentry RW, Kim JW (2008) Effective saturated hydraulic conductivity of two-dimensional random multifractal fields. Water resour. res. 44.W08410, doi: 10.1029/1007WR006199.

[65] Oleschko K, Korvin G, Munoz A, Velazquez J, Miranda ME, Carreon D, Flores L, Martinez M, Velasquez-Valle M, Brambila F, Parrot JF, Ronquillo G (2008) Mapping soil fractal dimension in agricultural fields with GPR. Nonlin. proc. geophy. 15: 711–725.

[66] Liu J, Ma X, Fu Q, Zhang Z (2012a) Joint multifractal of relationship between spatial variability of soil properties in different soil layers. Trans. chinese soc. agril. mech. 43: 37–42.

[67] Zeleke TB, Si BC (2007) Wavelet-based multifractal analysis of field scale variability in soil water retention. Water resour. res. 43. W07446, doi:10.1029/2006WR004957.

[68] Guo L, Li Y, Li M, Ren X, Liu C, Zhu D (2011) Multifractal study on spatial variability of soil hydraulic properties of Lou soil. Trans. chinese soc. agril. mech. 42: 50–58.

[69] Wang ZY, Shu QS, Xie LY, Liu ZX, Si BC (2011) Joint multifractal analysis of scaling relationships between soil water-retention parameters and soil texture. Pedosphere 21: 373–379.

[70] Liu J, Ma X, Zhang Z (2010) Spatial variability of soil water retention curve in different soil layers and its affecting factors. Trans. chinese soc. agril. mech. 41: 46–52.

[71] Liu J, Ma X, Zhang Z, Fu Q (2012) Pedotransfer functions of soil water retention curve based on joint multifractal. Trans. chinese soc. agril. mech. 43: 51–56.

[72] van Genuchten MT (1980) A closed-form equation for predicting the hydraulic conductivity of unsaturated soils. Soil sci. soc. am. j. 44:892–898.

[73] Kim G, Barros AP (2002) Downscaling of remotely sensed soil moisture with a modified fractal interpolation method using contraction mapping and ancillary data. Remote sens. environ. 83: 400–413.

[74] Oldak A, Pachepsky Y, Jackson TJ, Rawls WJ (2002) Statistical properties of soil moisture images revisited. J. hydrol. 255: 12–24.

[75] Lovejoy S, Tarquis AM, Gaonac'h H, Schertzer D (2008) Single- and multiscale remote sensing techniques, multifractals, and MODIS-derived vegetation and soil moisture. Vadose zone j. 7: 533–546.

[76] Mascaro G, Vivoni ER (2010) Statistical and scaling properties of remotely-sensed soil moisture in two contrasting domains in the North American monsoon region. J. arid environ. 74: 572–578.

[77] Mascaro G, Vivoni ER (2012) Comparison of statistical and multifractal properties of soil moisture and brightness temperature from ESTAR and PSR during SGP99. IEEE geosci. remote sens. lett. 9: 373–377.

[78] Mascaro G, Vivoni ER, Deidda R (2010) Downscaling soil moisture in the southern Great Plains through a calibrated multifractal model for land surface modeling applications. Water resour. res. 46. W08546, doi:10.1029/2009WR008855

[79] Mascaro G, Vivoni ER, Deidda R (2011) Soil moisture downscaling across climate regions and its emergent properties. J. geophys. res. -atm. 116. D22114, doi:10.1029/2011JD016231.

[80] Vazquez EV, Moreno RG, Miranda JGV, Diaz MC, Requejo AS, Ferreiro JP, Tarquis AM (2008) Assessing soil surface roughness decay during simulated rainfall by multifractal analysis. Nonlin. proc. geophy. 15: 457–468.

[81] Vazquez EV, Miranda JGV, Paz-Ferreiro J (2010) A multifractal approach to characterize cumulative rainfall and tillage effects on soil surface micro-topography and to predict depression storage. Biogeosciences 7: 2989–3004.

[82] Moreno RG, Alvarez MCD, Requejo AS, Tarquis AM (2008) Multifractal analysis of soil surface roughness. Vadose zone j. 7: 512–520.

[83] Roisin CJC (2007) A multifractal approach for assessing the structural state of tilled soils. Soil sci. soc. Am. j. 71: 15–25.

[84] San Jose Martinez F, Caniego J, Guber A, Pachepsky Y, Reyes M (2009) Multifractal modeling of soil microtopography with multiple transects data. Ecol. complex. 6: 240–245.

[85] Garcia-Moreno R, Diaz-Alvarez MC, Saa-Requejo A, Valencia-Delfa, JL (2011) Multiscaling analysis of soil drop roughness. In: Ozkaraova Gungor BE, editor, Principles, Application and Assessment in Soil Science, InTech, Slavka Krautzeka, Croatia.

[86] Caniego J, Ibanez JJ, Martinez FSJ (2006) Selfsimilarity of pedotaxa distributions at the planetary scale: A multifractal approach. Geoderma 134: 306–317

[87] Caniego FJ, Ibanez JJ, Martinez FSJ (2007) Renyi dimensions and pedodiversity indices of the earth pedotaxa distribution. Nonlin. proc. geophy. 14: 547–555.

[88] Zhao P, Shao M-A, Wang TJ (2011) Multifractal analysais of particle-size distributions of alluvial soils in the dam farmland on the Loess Plateau of China. Afr. j. agril. res. 6: 4177–4184.

[89] Grout H, Tarquis AM, Wiesner MR (1998) Multifractal analysis of particle size distributions in soil. Environ. sci. tech. 32: 1176–1182.

[90] Wang D, Fu B, Zhao W, Hu H, Wang Y (2008) Multifractal characteristics of soil particle size distribution under different land-use types on the Loess Plateau, China. Catena 72: 29–36.

[91] Paz-Ferreiro J, Vidal Vazquez E, Miranda JGV (2010) Assessing soil particle-size distribution on experimental plots with similar texture under different management systems using multifractal parameters. Geoderma 160: 47–56.

[92] Liu X, Zhang G, Heathman GC, Wang Y, Huang CH (2009) Fractal features of soil particle-size distribution as affected by plant communities in the forested region of Mountain Yimeng, China. Geoderma 154: 123–130.

[93] Vazquez EV, Garcia Moreno R, Tarquis AM, Requejo AS, Miras-Avalos JM, Paz-Ferreiro J (2010) Multifractal characterization of pore size distributions measured by mercury intrusion porosimetry. Proceedings of the 19th World Congress of Soil Science: Soil solutions for a changing world, Brisbane, Australia, 1–6 August 2010. Symposium 2.1.2 the physics of soil pore structure dynamics: pp. 57–60.

[94] Miranda JGV, Montero E, Alves MC, Gonzalez AP, Vazquez EV (2006) Multifractal characterization of saprolite particle-size distributions after topsoil removal. Geoderma 134: 373–385.

[95] Millan H, Gonzalez-Posada M, Aguilar M, Dominguez J, Cespedes L (2003) On the fractal scaling of soil data. Particle-size distributions. Geoderma 117: 117–128.

[96] Posadas AND, Gimenez D, Quiroz R, Protz R (2003) Multifractal characterization of soil pore systems. Soil sci. soc. am. j. 67: 1361–1369.

[97] Zhou H, Perfect E, Li BG, Lu YZ (2010) Effects of bit depth on the multifractal analysis of grayscale images. Fractals - complex geo. pattern. scale. nature soc. 18: 127–138.

[98] Zhou H, Perfect E, Lu YZ, Li BG, Peng XH (2011) Multifractal analyses of grayscale and binary soil thin section images. Fractals - complex geo. pattern. scale. nature soc. 19: 299–309.

[99] Martinez FSJ, Caniego FJ, Garcia-Gutierrez C, Espejo R (2007) Representative elementary area for multifractal analysis of soil porosity using entropy dimension. Nonlin. proc. geophy. 14: 503–511.

[100] Posadas A, Quiroz R, Tannus A, Crestana S, Vaz CM (2009) Characterizing water fingering phenomena in soils using magnetic resonance imaging and multifractal theory. Nonlin. proc. geophy. 16: 159–168.

[101] San Jose Martinez F, Caniego J, Guber A, Pachepsky Y, Reyes M (2009) Multifractal modeling of soil microtopography with multiple transects data. Ecol. complex. 6: 240–245.

[102] San JoseMartinez F, Martin MA, Caniego FJ, Tuller M, Guber A, Pachepsky Y, Garcia-Gutierrez C (2010) Multifractal analysis of discretized X-ray CT images for the characterization of soil macropores. Geoderma 156: 32–42.

[103] Jimenez-Hornero FJ, Gutierrez de Rave E, Giraldez JV, Laguna AM (2009) The influence of the geometry of idealised porous media on the simulated flow velocity: A multifractal description. Geoderma 150: 196–201.

[104] Olsson J, Persson M, Albergel J, Berndtsson R, Zante P, Ohrstrom P, Slah N (2002) Multiscaling analysis and random cascade modeling of dye infiltration. Water resour. res. 38.

[105] Green TR, Dunn GH, Erskine RH, Salas JD, Ahuja LR (2009) Fractal Analyses of Steady Infiltration and Terrain on an Undulating Agricultural Field. Vadose zone j. 8: 310–320.

[106] Tarquis AM, McInnes KJ, Key JR, Saa A, Garcia MR, Diaz MC (2006) Multiscaling analysis in a structured clay soil using 2D images. J. hydrol. 322: 236–246.

[107] Zhou X. (2004) Fractal and multifractal analysis of runoff time series and stream networks in agricultural watersheds. PhD Thesis submitted to the Virginia Polytechnic Institute and State University, Virginia, USA

[108] Meng H, Salas JD, Green TR, Ahuja LR (2006) Scaling analysis of space-time infiltration based on the universal multifractal model. J. hydrol. 322: 220–235.

[109] Perfect E, Gentry RW, Sukop MC, Lawson JE (2006) Multifractal Sierpinski carpets: Theory and application to upscaling effective saturated hydraulic conductivity. Geoderma 134: 240–252.

[110] Vidal-Vazquez E, Paz-Ferreiro J (2012) Multifractal characteristics of nitrogen adsorption isotherms from tropical soils. Soil sci. 177: 120–130.

[111] Grubich AO (2012) Multifractal structure of the Cs-137 fallout at small spatial scales. J. environ. radio. 107: 51–55.

[112] Wang D, Fu B, Lu K, Xiao L, Zhang Y, Feng X (2010) Multifractal analysis of land use pattern in space and time: A case study in the Loess Plateau of China. Ecol. complex. 7: 487–493.

[113] Banerjee S, He Y, Guo X, Si BC (2011) Spatial relationships between leaf area index and topographic factors in a semiarid grassland: Joint multifractal analysis. Aust. j crop sci. 5: 756–763.

[114] Eghball B, Schepers JS, Negahban M, Schlemmer MR (2003) Spatial and temporal variability of soil nitrate and corn yield: Multifractal analysis. Agro. j. 95: 339–346.

[115] Kravchenko AN (2008) Stochastic simulations of spatial variability based on multifractal characteristics. Vadose zone j. 7: 521–524.

[116] Pinuela JA, Andina D, McInnes KJ, Tarquis AM (2007) Wavelet analysis in a structured clay soil using 2-D images. Nonlin. proc. geophy. 14: 425–434.

[117] Pinuela J, Alvarez A, Andina D, Heck RJ, Tarquis AM (2009) Quantifying a soil pore distribution from 3D images: Multifractal spectrum through wavelet approach. Geoderma 155: 203–210.

[118] Pinuela JA, Andina D, Torres J, Tarquis AM (2009) Quantifying flow paths in clay soils using multifractal dimension and wavelet-based local singularity. Intell. auto. soft comput. 15: 605–617.

[119] Wang Y, Wu C, Pan Z (2011) Multifractal detrending moving-average analysis on the US Dollar exchange rates. Physica A – stat. mech. its appl. 390: 3512–3523.

[120] Jiang ZQ, Zhou WX (2011) Multifractal detrending moving-average cross-correlation analysis. Physical Review E 84. 016106, doi:10.1103/PhysRevE.84.016106.

Chapter 7

Fractal and Chaos in Exploration Geophysics

Sid-Ali Ouadfeul, Leila Aliouane and Amar Boudella

Additional information is available at the end of the chapter

1. Introduction

The fractal analysis has been widely used in exploration geophysics. In gravity and magnetism it is used for causative sources characterization [1,2, 3, 4]. In seismology, the fractal analysis is used for earthquake characterization [5, 6; 7]. In petrophysics the fractal analysis is ued for lithofacies classification and reservoir characterization [8, 9]. We cite for example the paper of Lozada-Zumaeta et al [10], they have analyzed the distribution of petrophysical properties for sandy-clayey reservoirs by fractal interpolation. San José Martínez et al [11] have published a paper for the detection of representative elementary area for multifractal analysis of soil porosity using entropy dimension.

The current chapter is composed of three application of the fractal analysis in geophysics. The first one consists to use the fractal analysis for facies identification from seismic data, the proposed idea is based on the estimation of the generalized fractal dimensions. After that an application to the pilot KTB borehole is realized The second part of this chapter is to apply the same technique but on the well-logs data for another objectives. This last, consist to identify heterogeneities. Application on the Pilot KTB borehole shows a robustness of this last. The third application of the fractal formalism for heterogeneities analysis from synthetic amplitude versus offset (AVO) data.

2. Facies recognition from seismic data using the fractal analysis

The fractal analysis has been widely used for seismic data processing, we cite for example the paper of Rivastava and Sen [11]. They have developed a new fractal-based stochastic inversion of poststack seismic data using very fast simulated annealing. A Simultaneous stochastic inversion of prestack seismic data using hybrid evolutionary algorithm based on the fractal process has been developed recently by Saraswat and Sen [12].

Facies recognition from seismic data using the fractal analysis has becoming a very interesting subject of research; in fact many papers have been published in this topic.

We cite for example the paper of Lopez and Aldana [13], these last have used the wavelet based fractal analysis for facies recognition and waveform classifier from

Oritupano-A field.

Barnes [14] has used the fractal analysis of Fault Attributes Derived from Seismic Discontinuity Data. Nath and Dewangan [15], have established a technique to detect reflections from seismic attributes. Gholamy et al [16], have proposed a technique based on frcatal methods for automatic detection of interfering seismic wavelets.

Here, we present a wavelet-based fractal analysis technique for reflection recognition from seismic data; we used the so-called wavelet transform modulus maxima lines, to calculate the generalized fractal dimensions. We start by describing the principles of the continuous wavelet transform and the WTMM. The next section consists to explain the principle of the seismic data processing algorithm. The proposed idea is applied to the synthetic seismic seismogram of the Germann KTB pilot borehole data. We finalize the paper by a conclusions and a discussion of the obtained results.

2.1. The Wavelet Transform Modulus Maxima lines (WTMM) method

The wavelet transform modulus maxima lines (WTMM) is a multifractal formalism proposed by Arneodo et al [17], the WTMM is based on the continuous wavelet transform. Lets us define an analyzing wavelet $\psi(z)$, the continuous wavelet transform of a signal S(z) is given by [18]:

$$C_s(a,b) = \frac{1}{\sqrt{a}} \int_{-\infty}^{+\infty} s(z)\psi^*(\frac{z-b}{a})dz \tag{1}$$

Where a: is a scale parameter.

$\psi(\frac{z-b}{a})$ Is the dilated version of the analyzing wavelet.

$\psi^*(z)$ is the conjugate of $\psi(z)$

The analyzing wavelet must check the admissibility condition:

$\int_{-\infty}^{+\infty} \psi(z)dz = 0$ in frequency domain this condition is equivalent to :

$$\int_{-\infty}^{+\infty} \frac{\psi(w)}{\omega}dw = 0 \tag{2}$$

For some data processing requirements $\psi(z)$ must have N vanishing moments:

$\int_{-\infty}^{+\infty} z^n\psi(z)dz = 0$ for all n, 0≤n≤N-1

The first step on the WTMM consists to calculate the modulus of the continuous wavelet transform, after that the maxima of this modulus are calculated. We call maxima of the modulus of the CWT at point b_0 if for all $b \to b_0$ $|C_s(a, b_0)| > |C_s(a, b)|$.

The next step is to calculate the function of partition $Z(q, a)$. If one call $L(b)$ the set of maxima of the modulus of the CWT, the function of partition is defined by :

$$Z(q, a) = \sum_{L(b)} \left| C_s(a, b) \right|^q \tag{3}$$

Where q is the moment order.

The spectrum of exponents $\tau(q)$ is related to the function of partition for law scales by :

$$Z(q, a) = a^{\tau(q)} \text{ if } a \to 0 \tag{4}$$

Note that the spectrum of exponents is estimated by a simple linear regression of $\log(Z(q, a))$ versus $\log(a)$

The generalized fractal dimension is given by [17]:

$$D_q = \frac{\tau(q)}{(q-1)} \tag{5}$$

One can distinguish three important values of D_q:

If q=0: the definition becomes the capacity dimension.

If q=1: the definition becomes the information dimension.

If q=2: the definition becomes the correlation dimension.

2.2. The processing algorithm

The proposed idea consists to apply the so-called the wavelet transform modulus maxima lines (WTMM) method with a moving window of 128 samples, the window center will be moved at each 64 samples of the seismic trace. Note that 128 is the less required number of samples for the convergence of the function of partition. The analyzing wavelet is the complex Morlet [18]. After that, the three generalized fractal dimensions that consist to q=1, 2 and 3 are calculated. The goal is to show that these fractal dimensions can be used for thin bed and facies identification and eliminate the noise effect that can give false stratigraphic formations. The detailed flow chart of the proposed idea is presented in figure 1.

2.3. Application to KTB synthetic seismic data

To check the efficiency of the proposed technique, we have analyzed the Kontinentales Tiefbohrprogramm de Bundesreplik Deutschland (KTB) borehole synthetic seismic seismogram calculated from the well-logs data.

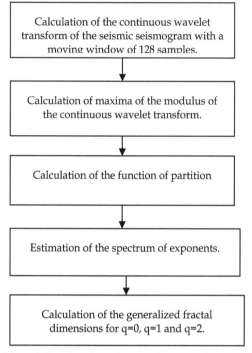

Figure 1. Flow chart of the multifractal analysis of seismic data using the generalized fractal dimensions

2.3.1. Geological context

The planning for the KTB super-deep borehole project, known as German Continental Deep Drilling Program, began in 1978 and drilling started in 1987 on the test hole KTB-VB (finished 1989) and the main borehole KTB-HB started in 1990 (finished 1994). In only 4 years KTB-HB was drilled down to a depth of 9101 m (29,859 ft). They made a number of very surprising finds and non finds including water deep below the surface.

Geologists had formed a picture of the crust at the Windischeschenbac site by examining rock outcrops and tow dimensional (2D) seismic measurements (see figure 2).

The lithology model shows an alternating of paragneisses, metabasite, amphibolites and hornblende gneiss and alternating layers of gneisses and amphibolites.

The structural model shows fomrtaions dipping 50° and 75° south-southwest over the first 3000m, followed by a rotation of the dip to the east with a much shallower dip of 25° in the fold hinge. These models were built from interpretation of core data and boreholes images such as Fomration MicroScanner images. The formation appears to have twisted and pilled up [19].

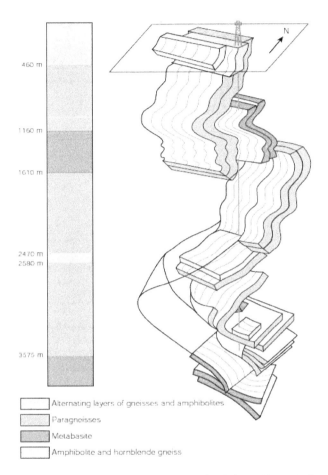

Figure 2. Simplified lithology and structure revealed by KTB borehole. (a) Alternating layers of metamorphic rocks. (b) Structural model shows formations dipping.

2.3.2. Data processing

The proposed idea has been applied on the KTB synthetic seismogram, figures 3a and 3b present the Sonic Velocity of the Primary wave and the formation density, after that the reflectivity function at normal incidence is calculated, for this last equation 06 is used [24].

$$Cr_i = \frac{\rho_{i+1}Vp_{i+1} - \rho_i Vp_i}{\rho_{i+1}Vp_{i+1} + \rho_i Vp_i} \tag{6}$$

Where: ρ_i is the formation density at depth Z_i

Vp_i is the Velocity of the P wave depth Z_i

Figure 3 shows this reflectivity versus the depth, note that the processed depth interval is [290m, 4000m], with a sampling interval of 0.1524m.

The synthetic seismic seismogram is than calculated using the convolution model, in fact a seismic trace with an emitted source wavelet $W(t)$ is given by [24]:

$$T(t) = W(t) * Cr(t) + N(t) \tag{7}$$

N(t) is a noise.

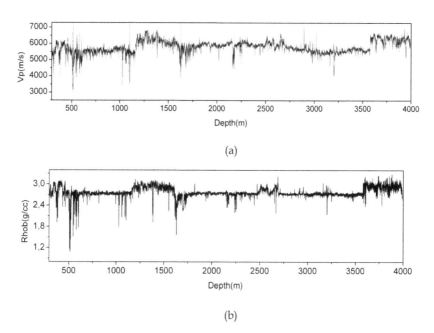

(a)

(b)

Figure 3. (a)Velocity of the P wave of the pilot KTB borehole versus the depth, (b)Density log of the pilot KTB borehole versus the depth

The emitted wavelet used in this paper is the Ricker, figure 5 shows the graph of this last versus the time. Figure 6 shows the synthetic seismic seismogram calculated using the convolution model. One can remark that is very easy to identify facies limits in this trace.

To check the robustness of the generalized fractal dimensions, the seismic seismogram is now noised with 200% of white noise.

Figure 7 shows the noisy seismic seismogram, on can remark that we are not able to identify the facies limit, for the full depth interval, for example geological formations limited between 2750m and 3000m are hidden by noise. The noisy seismic seismogram is then processed using the proposed algorithm. Figure 8 shows the three fractal dimensions versus the depth compared with the geological map.

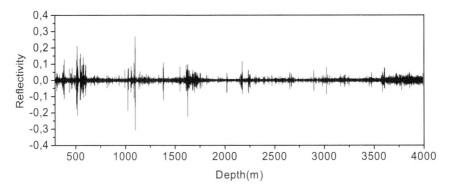

Figure 4. Normal incidence reflectivity versus the depth of the pilot KTB borehole.

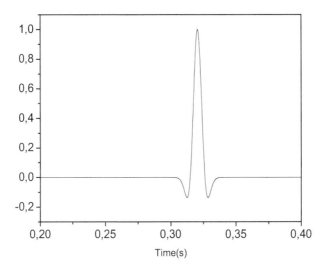

Figure 5. Graph of the Ricker wavelet versus the time.

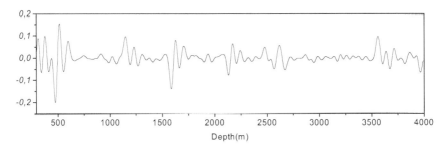

Figure 6. Synthetic seismic seismogram of the pilot KTB borehole.

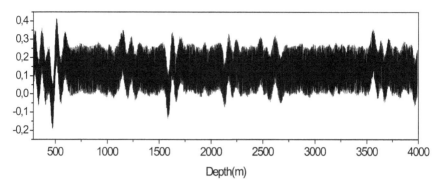

Figure 7. Noisy seismic seismogram

2.4. Results interpretation and conclusion

One can remark that the principles contacts that exist in geological cross section over the drill are identified by the generalized fractal dimensions (see blue dashed lines in figure 8), however the fractal dimension D_0 (called the capacity dimension) is not always very robust than the information and the correlation dimensions. For example at the depth 460m (first blue dashed line) D_0 is not able to detect this contrast. Otherwise, the generalized fractal dimensions are very robust tools for identification of amplitudes that are due to the lithology variation. For example in the depth interval [2750m, 3000m] the three fractal dimensions have identified many contacts that are totally hidden by noise. The multifractal analysis of seismic data can be used for facies identification, we suggest introducing this last in seismic data processing flow and software.

3. Heterogeneities analysis from well-logs data using the multifractal analysis and the continuous wavelet transform

The fractal analysis has been widely used for heterogeneity analysis from well-llogs data. We cite for example the paper of Kue et al [20]. It shows how texture logs computed from multifractal analysis of dipmeter microresistivity signals can be used for characterizing lithofacies in combination with conventional well logs. Li [21], has analyzed the well-logs data as a Fractional Brownian Motion (fBm) model, he has examined the paradoxical results found in the literature concerning the Hurst exponents. Ouadfeul and Aliouane [08] have published a paper that uses the multifractal analysis for lithofacies segmentation from well-logs data. In this paper, we use the wavelet-based generalized fractal dimensions for heterogeneities analysis. The proposed idea has been applied to a synthetic and real sonic well-logs data of the Kontinentales Tiefbohrprogramm de Bundesreplik Deutschland (KTB) also known as German Continental Deep Drilling Program. We start the paper by describing a wavelet-based multifractal analysis called the Wavelet Tranform Modulus Maxima lines (WTMM) method, after that the processing algorithm of heterogeneities analysis from well-logs data is well detailed. The proposed idea is then applied to the synthetic and real data of the Pilot and the Main KTB boreholes. We finalize the paper by the results discussion and a conclusion.

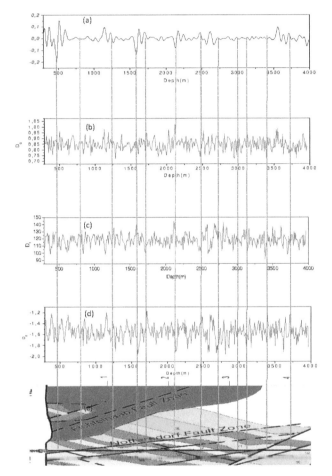

Figure 8. Multifractal Analysis of the noisy seismic seismogram: (a) Seismic seismogram without noise. (b) D0. (c) D1. (d) D2. (e) Geological cross section over the pilot KTB drilling.

3.1. The processing algorithm

The processing algorithm is based on the application of the Wavelet Transform Modulus Maxima lines (WTMM) method at each 128 samples of the signal. A moving window with this last number of samples is used. The window center is moved by 64 samples. At each window the three generalized fractal dimensions D0, D1 and D2 are calculated. Note that

128 samples is the less number of samples where the WTMM can be applied [23]. The analyzing wavelet is the Compex Morlet defined by equation 6:

$$\psi(t) = \exp(i\omega t)\exp(-t^2/2) - \sqrt{2}\exp(-\Omega^2/4)\exp(i\omega t)\exp(-t^2). \tag{8}$$

Ouadfeul and Aliouane [8,9], have showed that the optimal value of Ω for a better estimation of the Hurst exponent is equal to 4.8.

The purpose of this work is to use the generalized fractal dimensions for lithology segmentation and heterogeneities analysis.

3.2. Application to synthetic data

To check the efficiency of the generalized fractal dimensions calculated using the WTMM formalism for boundaries layers delimitation, we have tested this last at a synthetic model formed with four roughness coefficients to model the geological variation of lithology. Each lithology will give a synthetic well-log considered as a fractional Brownian motion (fBm) model. Parameters of the synthetic model are detailed in table 1. Figure 9 shows the generated synthetic model. The generalized fractal dimensions obtained by WTMM analysis of this synthetic well-log are presented in figure 10. We can observe that the fractal dimensions D_1 and D_2 are able to detect exactly the boundaries of each layer; however the capacity dimension D_0 is not sensitive to lithology variation. By consequence, D_0 cannot be used for lithology segmentation.

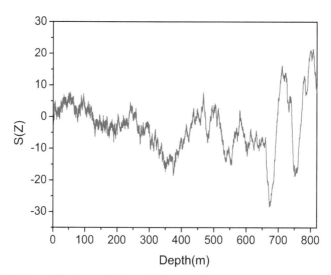

Figure 9. Synthetic well-log data realization with four roughnesses

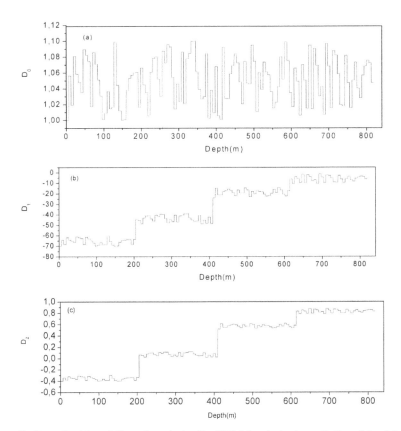

Figure 10. Generalized fractal dimensions obtained by WTMM analysis of a synthetic well-log data. (a) : D0, (b): D1 (c): D2

3.3. Application to KTB boreholes data

The proposed idea is applied on the sonic Primary (P)) wave velocity well-log of the pilot KTB borehole. The goal is to check its efficiency on real well logs data.

3.3.1. Data processing

Figures 3a show the Primary (P) wave velocity in the Pilot KTB borehole. In this paper, only the depth interval [290m, 4000m] is considered.

Data are recorded with a sampling interval of 0.1524m. The wavelet transform modulus maxima lines method is applied with a sliding window of 128 samples. The window center is moved with 64 samples. The analyzing wavelet is the Complex Morlet (see equation 6). At each window the spectrum of exponents is estimated and the three fractal dimensions are calculated using equation 5. Figure11 shows the three fractal dimensions obtained by multifractal analysis of the Primary wave velocity for the Pilot KTB borehole.

Layer number	01	02	03	04
Thickness (m)	204.80	204.80	204.80	204.80
Roughness Coefficient	0.33	0.51	0.76	0.89
Number of Samples	2048	2048	2048	2048

Table 1. Parameters of the synthetic well-log data.

3.3.2. Results interpretation and conclusion

Obtained results are compared with the simplified lithology and geological cross section of the area (figure 11). One can remark easily that the capacity dimensions D_0 is not sensitive to the lithology variation. Segmentation models based on D_1 and D_2 are made. It is clear that the information and correlation dimensions are able to detect lithological transitions that are defined by geologists (Black dashed lines in figure 11). Geological model based on the fractal dimensions is proposed (see blue dashed lines in the same figures).

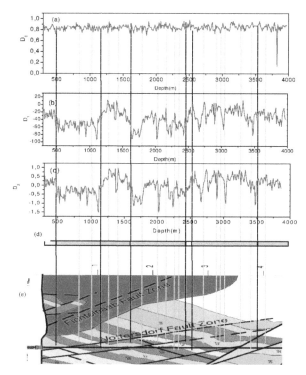

Figure 11. Generalized fractal dimensions compared with the Geological Cross-Section for the pilot KTB borehole. The analyzed log is the velocity of the P wave.

We have implanted a technique of lithology segmentation based on the generalized fractal dimensions. The proposed idea is successfully applied on the pilot KTB borehole. We recommend application of the proposed method on the potential magnetic and gravity data for contact identification and causative sources characterization. The proposed idea can be applied for reservoir characterization and lithofacies segmentation from well-logs data; it can help for oil exploration and increasing oil recovery.

4. Fractal analysis of 2D seismic data for heterogeneities analysis

The 1D wavelet transform modulus maxima lines WTMM is a multifractal analysis technique based on the summation of the modulus of the continuous wavelet transform(CWT) on its maxima. The obtained function is used to estimate two spectrums, one is the spectrum of exponents and another is the spectrum of singularities. The WTMM was used in various domains to resolve many scientific problems [24][25]. A generalized wavelet transform modulus maxima lines WTMM in the 2D domain are used by many researchers to establish physicals problems [25][26]. The wavelet transform has been applied in seismic image processing; Miao and Moon [27] have published a paper on the analysis of seismic data using the wavelet transform. A New sparse representation of seismic data using adaptive easy-path wavelet transform has been developed by Jianwei et al [28].

The continuous wavelet transform, is used by Pitas et al [29] for texture analysis and segmentation of seismic images. In this paper we process the intercept attribute of 3D synthetic seismic AVO data by the 2D WTMM to establish the problem of heterogeneities. It is very complex and need advanced processing tools to get more ideas about morphology of rocks.

4.1. Brownian fractional motion and synthetic model

For $H \in (0, 1)$, a Gaussian process $\{B_H(t)\}_{t \geq 0}$ is a fractional Brownian Motion if for all $t, s \in \Re$ it has [30] :

1. A Mean: $E[B_H(t)] = 0$
2. A Covariance: $E[B_H(t)B_H(s)] = (1/2) \{|t|^{2H} + |s|^{2H} - |t - s|^{2H}\}$

H is the Hurst exponent [31].

We suppose now that we have a geological model of two layers the first is homogonous with the following physical parameters:

1. $Vp \approx 3500 \left(\dfrac{m}{s} \right)$

2. The density is calculated using the Gardner model [32]: $\rho \approx 1.741 \times VP^{0.25} = 2.38 \left(\dfrac{g}{cc} \right)$

3. The velocity of the shear wave is estimated using Castagna Mud-rock line [33]: $Vs = 0.8621 X Vp - 1172.4 = 1744.95 (m / s)$

The second is a heterogeneous model with the parameters detailed below.

We suppose that the velocity of the P wave, the velocity of the shear wave and the density of the synthetic model are a Brownian fractional motion model versus the azimuth θ. $0 \le \theta \prec 180$

A generation of the three geological parameters is represented in figures 12.a, 12.b, and 12.c.

We suppose that the three mechanical parameters are in the following range of intervals:

$$2000 \prec Vp \le 6000$$
$$1000 \le Vs \le 4000$$
$$2.1 \le \rho \le 3.$$

Where Vp(m/s) is the velocity of the P wave, Vs(m/s) is the velocity of the shear wave and $\rho(g / cc)$ is the density.

We calculated the reflection coefficients at the null offset R0, which are depending to the azimuth or to X and Y coordinates, obtained results are represented in figure 13.

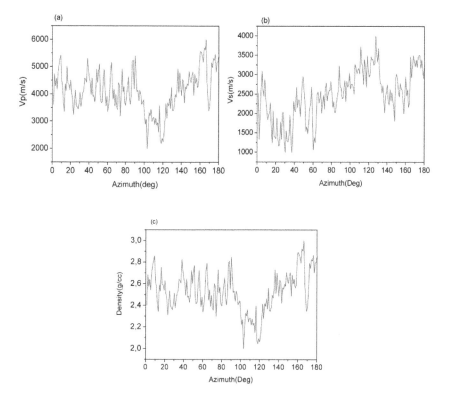

Figure 12. Physical parameters of a synthetic heterogonous layer generated randomly versus azimuth (a) Velocity of the P wave (b) Velocity of the S wave (c) Density of the model.

Figure 13. Response of a model of two layers the first is homogeneous and the second is heterogeneous

4.2. The 2D wavelets transform modulus maxima lines WTMM

The 2D wavelet transform modulus maxima lines WTMM is a signal processing technique introduced by Arneodo and his collaborators in image processing [26][27] .It was applied in medical domain as a tool to detect cancer of mammograms and in image processing [25]. The flow chart of this method is detailed in figure 14.

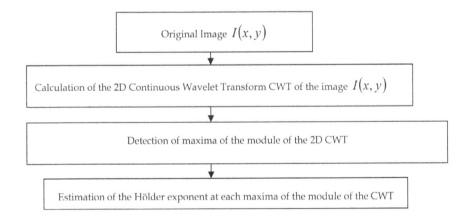

Figure 14. Flow chart of the 2D wavelet transform modulus maxima lines WTMM

4.3. Application to synthetic model

We applied the proposed technique at the synthetic model proposed above. Figure 15 presents the modulus of the wavelet coefficients at the lower scale a=2.82m. Figure 16 presents the phase proposed by Arneodo et al [31]; the Skelton of the modulus of the continuous wavelet transform is presented in figure 17 and the local Hölder exponents estimated at each point of maxima is presented in figure 18 [31]. One can remark that the Hölder exponents map can provide information about reflection coefficient behavior. So it can be used as a new seismic attribute for lithology analysis and heterogeneities interpretation.

Figure 15. Modulus of the wavelet transform plotted at the low dilatation

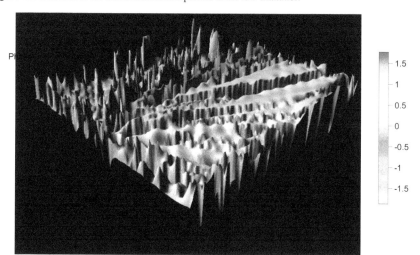

Figure 16. Phase of the 2D continuous wavelet transform

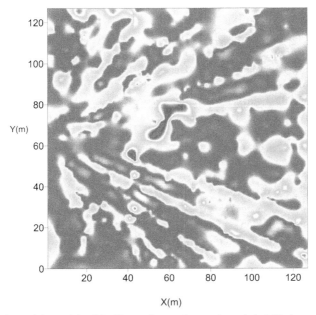

Figure 17. Skeleton of the module of the 2D wavelet transform at the scale (a=2.82m)

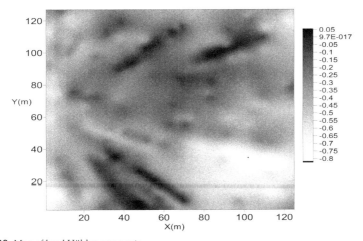

Figure 18. Map of local Hölder exponents

4.4. Results interpretation and conclusion

Analysis of the obtained results shows that the 2D WTMM analysis can enhance seismic data interpretation. Hölder exponents map (figure 18) is a good candidate for reservoir heterogeneities analysis. This map is an indicator of geological media roughness.

Conclusion

Application of the proposed technique to an AVO synthetic heterogeneous model shows that the 2D wavelet transform modulus maxima lines (WTMM) is able to give more information about reservoirs rock heterogeneities. The local Hölder exponent can be used as a supplementary seismic attribute to analyze reservoir heterogeneities. The proposed analysis can help the hydrocarbon trapping research and analysis, fracture detection and fractured reservoirs analysis. We suggest application of the proposed philosophy at real seismic AVO data and its attribute, for example the Intercept, The gradient, the Fluid Factor proposed by Smith and Gildow [34] and the attribute product Intercept*Gradient .

5. Conclusion

As a conclusion of the current chapter, we can say that the fractal analysis has showed it's powerful to help exploration geophysics. Application to synthetic and real seismic and well-logs data shows clearly that this analysis of the earth response considered as a chaotic system can help geosciences.

Author details

Sid-Ali Ouadfeul
Algerian Petroleum Institute, IAP, Algeria
Geophysics Department, FSTGAT, USTHB, Algeria

Leila Aliouane
Geophysics Department, FSTGAT, USTHB, Algeria
Geophysics Department, LABOPHYT, FHC, UMBB, Algeria

Amar Boudella
Geophysics Department, LABOPHYT, FHC, UMBB, Algeria

6. References

[1] Maus, S, Dimri, VP., 1994, Fractal properties of potential fields caused by fractal sources. Geophys Res Lett 21: 891-894.

[2] Maus, S. and Dimri V.P., 1996, Depth estimation from the scaling power spectrum of potential field Geophysical J. Int., 124,113-120.

[3] Fedi, M., 2003, Global and local multiscale analysis of magnetic susceptibility data. Pure Appl Geophys 160:2399-2417

[4] Fedi, M., Quarta, T., Santis. AD., 1997, Inherent power-law behavior of magnetic field power spectra from a Spector and Grant ensemble. Geophysics 62:1143-1150.

[5] Ravi prakash, M. and V.P., Dimri, 2000, Distribution of the aftershock sequence of Latur earthquake in time and space by fractal approach, JGSI, 55, 167-174.

[6]   Teotia, S. S. and Kumar, D., 2011, Role of multifractal analysis in understanding the preparation zone for large size earthquake in the North-Western Himalaya region, Nonlin. Processes Geophys., 18, 111-118, doi:10.5194/npg-18-111-2011,

[7]   Dimri VP, Nimisha V, Chattopadhyay S, 2005, Fractal analysis of aftershock se-quence of Bhuj earthquake - a wavelet based approach. Curr Sci.

[8]   Ouadfeul, S, and Aliouane, L., 2011, Multifractal analysis revisited by the continuous wavelet transform applied in lithofacies segmentation from well-logs data, International Journal of Applied Physics and Mathematics vol. 1, no. 1, pp. 10-18, 2011.

[9]   Ouadfeul S, Aliouane, L., 2011, Automatic lithofacies segmentation suing the wavelet transform modulus maxima lines (WTMM) combined with the detrended fluctuations analysis (DFA), Arab Journal of Geosciences.

[10] Lozada-Zumaeta, M., Arizabalo, R. D., Ronquillo-Jarillo, G., Coconi-Morales, E., Rivera-Recillas, D., and Castrejón-Vácio, F., 2012, Distribution of petrophysical properties for sandy-clayey reservoirs by fractal interpolation, Nonlin. Processes Geophys., 19, 239-250, doi:10.5194/npg-19-239-2012.

[11] San José Martínez, F., Caniego, F. J., García-Gutiérrez, C., and Espejo, R., 2007, Representative elementary area for multifractal analysis of soil porosity using entropy dimension, Nonlin. Processes Geophys., 14, 503-511, doi:10.5194/npg-14-503.

[12] Srivastava, P.R, and Sen, M.K, 2009, Fractal-based stochastic inversion of poststack seismic data using very fast simulated annealing, J. Geophys. Eng. 6 412 doi:10.1088/1742-2132/6/4/009.

[13] López, M. and Aldana, M., 2007, Facies recognition using wavelet based fractal analysis and waveform classifier at the Oritupano-A Field, Venezuela, Nonlin. Processes Geophys., 14, 325-335, doi:10.5194/npg-14-325-2007.

[14] Barnes, A.E, 2005, Fractal Analysis of Fault Attributes Derived from Seismic Discontinuity Data, 67th EAGE Conference & Exhibition, extended abstract.

[15] Nath, K., and Dewangan, P., 2002, Detection of seismic reflections from seismic attributes through fractal analysis, Geophysical Prospecting, Vol 50, Issue 3, pages 341-360.

[16] Gholamy, S., Javaherian, A., Ghods, A., 2008, Automatic detection of interfering seismic wavelets using fractal methods, J.Geophys.Eng.5(2008)338–347, doi:10.1088/1742-2132/5/3/009.

[17] Arneodo, A., Grasseau, G., and Holschneider, M., 1988. Wavelet transform of multifractals, Phys. Rev. Lett. 61:2281-2284.

[18] Grossman, A., Morlet, J-F., 1985, Decomposition of functions into wavelets of constant shape, and related transforms , in :Streit , L., ed., mathematics and physics ,lectures on recents results , World Scientific Publishing , Singaporemm .

[19] Bram, K., Draxler, J., Hirschmann, G., Zoth, G., Hiron, S., Kuhr, M., 1995, The KTB borehole –Germany's Superdeep Telescope into the Earth's Crust, OilField Review, January.

[20] Khue,P., Huseby, O., Saucier, A., and Muller, J., 2002, Application of generalized multifractal analysis for characterization of geological formations, J. Phys.: Condens. Matter 14 2347 doi: 10.1088/0953-8984/14/9/323.

[21] Li, C-F., 2003, Rescaled range and power spectrum analyses on well-logging data, Geophy. J. Int, , Vol.153, Issue 1, pp. 201-212.

[22] Biswas, A., Zeleke, T. B., and Si, B. C., 2012, Multifractal detrended fluctuation analysis in examining scaling properties of the spatial patterns of soil water storage, Nonlin. Processes Geophys., 19, 227-238, doi:10.5194/npg-19-227-2012.

[23] Ouadfeul , S., 2006 , Automatic lithofacies segmentation using the wavelet transform modulus maxima lines (WTMM) combined with the detrended fluctuation analysis (DFA), 17 International geophysical congress and exhibition of turkey , Ankara .

[24] Ouadfeul , S., 2007 , Very fines layers delimitation using the wavelet transfrom modulus maxima lines WTMM combined with the DWT , SEG SRW ,Antalya , Turkey.

[25] Kestener, P., 2003, Analyse multifractale 2D et 3D à l'aide de la transformée en ondelettes : Application en mammographie et en turbulence développée, Thèse de doctorat, Université de paris sud.

[26] Ouadfeul , S; Aliouane , L., 2010; Multiscale analysis of GPR data using the continuous wavelet transform, presented in GPR 2010, IEEE Xplore Compiliance, doi:10.1109/ICGPR.2010.5550177.

[27] Miao, X., and Moon, W., 1999, Application of wavelet transform in reflection seismic data analysis, Geosciences Journal , Volume 3, Number 3, 171-179, DOI: 10.1007/BF02910273.

[28] Jianwei Ma; Plonka, G.; Chauris, H.; A New Sparse Representation of Seismic Data Using Adaptive Easy-Path Wavelet Transform, Geoscience and Remote Sensing Letters, IEEE, Vol 7, Issue:3, pp540 – 544, doi10.1109/LGRS.2010.2041185

[29] Pitas, I.; Kotropoulos, C., 1989, Texture analysis and segmentation of seismic images International Conference on Acoustics, Speech, and Signal Processing, 1989. ICASSP-89., 1989, doi:10.1109/ICASSP.1989.266709.

[30] Peitgen, HO., Saupe, D., 1987, The science of Fractal Images. New York: Springer Verlag

[31] Arneodo, A., Decoster, N., Kestener, P and Roux, S.G., 2003, A wavelet-based method for multifractal image analysis: From theoretical concepts to experimental applications, Advances In Imaging And Electron Physics 126, 1--92.

[32] Gardner, G., Gardner, L., and Gregory, A., 1974, Formation velocity and density – the diagnostic basis for stratigraphic traps: Geophysics, 39, pp 770-780.

[33] Castagna, J., Batzle, M., and Eastwood, R., 1985, Relationships between compressional-wave and shear wave velocities in clastic silicate rocks: Geophysics, 50, pp 571-581.

[34] Smith, G.C. and Gildow, P.M., 1987, Weighted stacking for rock property estimation and detection of gas: Geophysical Prospecting, 35, 993-1014.

Chapter 8

Well-Logs Data Processing Using the Fractal Analysis and Neural Network

Leila Aliouane, Sid-Ali Ouadfeul and Amar Boudella

Additional information is available at the end of the chapter

1. Introduction

One of the main goals of geophysical studies is to apply suitable mathematical and statistical techniques to extract information about the subsurface properties. Well logs are largely used for characterizing reservoirs in sedimentary rocks. In fact it is one of the most important tools for hydrocarbon research for oil companies. Several parameters of the rocks can be analyzed and interpreted in term of lithology, porosity, density, resistivity, salinity and the quantity and the kind of fluids within the pores.

Geophysical well-logs often show a complex behavior which seems to suggest a fractal nature (Pilkington & Tudoeschuck, 1991; Wu et al., 1994; Turcotte, 1997; Ouadfeul, 2006; Ouadfeul and Aliouane 2011; Ouadfeul et al, 2012). They are geometrical objects exhibiting an irregular structure at any scale. In fact, classifying lithofacies boundary from borehole data is a complex and non-linear problem. This is due to the fact that several factors, such as pore fluid, effective pressure, fluid saturation, pore shape, etc. affect the well log signals and thereby limit the applicability of linear mathematical techniques. To classify lithofacies units, it is, therefore, necessary to search for a suitable non-linear method, which could evade these problems.

The scale invariance of properties has led to the well known concept of fractals (Mandelbrot, 1982). It is commonly observed that well log measurements exhibit scaling properties, and are usually described and modelled as fractional Brownian motions (Pilkington & Tudoeschuck, 1991; Wu et al. 1994; Kneib 1995; Bean, 1996; Holliger 1996; Turcotte 1997; Shiomi et al. 1997; Dolan et al 1998; Li 2003; Ouadfeul, 2006; Ouadfeul and Aliouane, 2011; Aliouane et al; 2011). In previous works (Ouadfeul, 2006; Ouadfeul and Aliouane, 2011; Aliouane et al, 2011), we have shown that well logs fluctuations in oil exploration display scaling behaviour that has been modelled as self affine fractal processes. They are therefore considered as fractional Brownian motion (fBm), characterized by a fractal k-β power spectrum model where k is the wavenumber and β is related to the Hurst parameter (Hermman,1997; Ouadfeul and Aliouane, 2011). These processes are monofractal whose complexity is defined by a single global coefficient, the Hurst parameter H, which is closely related to the Hölder degree regularity Thus, characterizing scaling behavior amounts to estimating some power law exponents.

Petrophysical properties and classification of lithofacies boundaries using the geophysical well log data is quite important for the oil exploration. Multivariate statistical methods such as principle component and cluster analyses and discriminate functions analysis have regularly been used for the study of borehole data. These techniques are, however, semi-automated and require a large amount of data, which are costly and not easily available every time.

The modern data modeling approach based on the Artificial Neural Network (ANN) techniques is inherently nonlinear and completely data-driven requiring no initial model and hence provide an effective alternative approach to deal with such a complex and non-linear geophysical problem. Some researchers have been engaged in classifying lithofacies units from the recorded well logs data. They have recently employed statistical and ANN methods (Aliouane et al, 2011).

In this work, we show that the fractal analysis is not able to improve lithofacies classification from well-logs data using the Self-Organizing Map neural Network. We analyze several petrophysical properties recorded in two boreholes, Well01and Well02 located in Berkine basin in the northeast of the Saharan platform (Algeria). This basin is considered as a vast Palaeozoic depression in which the crystalline basement is covered by an important sedimentary series. Lithologically, the explored geological unit at the drill site consists of four main facies units: clay, sandstone and alternations of clayey sandstone and Sandy Clay (Well Evaluation Conference., 2007).

A fractal model is assumed for the logs and they are analyzed by the Continuous Wavelet Transform (CWT) which maps the measured logs to profiles of Hölder exponents. We use the estimated wavelet Hölder exponents rather than the raw data measurements to check the classification process initiated by a self organizing map of Kohonen procedure.

In this chapter we first present a short mathematical description to show that the CWT is the suitable tool used to analyze concept of a self affine process. Second, we describe the neural network method, particularly, the Kohonen's Self Organizing Map (SOM) and its derived processing algorithm. Finally, we show the fractal analysis effect on the Self-Organizing Map neural network for lithofacies classification.

2. Wavelet analysis of scaling processes

Here we review some of the important properties of wavelets, without any attempt at being complete. What makes this transform special is that the set of basis functions, known as wavelets, are chosen to be well-localized (have compact support) both in space and frequency (Arneodo et al., 1988; Arneodo et al., 1995; Ouadfeul and Aliouane, 2010). Thus, one has some kind of "dual-localization" of the wavelets. This contrasts the situation met for the Fourier Transform where one only has "mono-localization", meaning that localization in both position and frequency simultaneously is not possible.

The CWT of a function s(z) is given by Grossmann and Morlet, (1985) as:

$$C_s(a,b) = \frac{1}{\sqrt{a}} \int_{-\infty}^{+\infty} s(z)\psi^*(z)dz,$$ (1)

Each family test function is derived from a single function $\psi(z)$ defined to as the analyzing wavelet according to (Torresiani, 1995):

$$\psi_{a,b}(z) = \psi(\frac{z-b}{a}),$$ (2)

Where $a \in R^{+*}$ is a scale parameter, $b \in R$ is the translation and ψ^* is the complex conjugate of ψ. The analyzing function $\psi(z)$ is generally chosen to be well localized in space (or time) and wavenumber. Usually, $\psi(z)$ is only required to be of zero mean, but for the particular purpose of multiscale analysis $\psi(z)$ is also required to be orthogonal to some low order polynomials, up to the degree n−1, i.e., to have n vanishing moments :

$$\int_{-\infty}^{+\infty} z^n \psi(z)dz = 0 \quad for \ 0 \le n \le p-1$$ (3)

According to equation (3), p order moment of the wavelet coefficients at scale a reproduce the scaling properties of the processes. Thus, while filtering out the trends, the wavelet transform reveals the local characteristics of a signal, and more precisely its singularities.

It can be shown that the wavelet transform can reveal the local characteristics of s at a point z0. More precisely, we have the following power-law relation (Hermann, 1997; Audit et al., 2002):

$$|C_s(a,z_0)| \approx a^{h(z0)} , \ whe \quad a \to 0^+$$ (4)

where h is the Hölder exponent (or singularity strength). The Hölder exponent can be understood as a global indicator of the local differentiability of a function s.

The scaling parameter (the so-called Hurst exponent) estimated when analysing process by using Fourier Transform (Ouadfeul and Aliouane, 2011) is a global measure of self-affine process, while the singularity strength h can be considered as a local version (i.e. it describes 'local similarities') of the Hurst exponent. In the case of monofractal signals, which are characterized by the same singularity strength everywhere (h(z) = constant), the Hurst exponent equals h. Depending on the value of h, the input signal could be long-range correlated (h > 0.5), uncorrelated (h = 0.5) or anticorrelated (h < 0.5).

3. Neural network method

The Artificial Neural Network (ANN) based approaches have proved to be one of the robust and cost-effective alternative means to successfully resolve the lithofacies boundaries from well log data (Gottlib-Zeh et al, 1999; Aliouane et al, 2011). The method has its inherent learning ability to map some relation between input and output space, even if there is no explicit a priori operator linking the measured lithofacies properties to the well log response.

3.1. Self Organizing map of kohonen

A Self Organizing neural network, or SOM, is a collection of n reference vectors organised in a neighbourhood network, and they have the same dimension as the input vectors (Kohonen, 1998). Neighbourhood function is usually given in terms of a two-dimensional neighbourhood matrix $\{W(i,j)\}$. In a two-dimensional map, each node has the same neighbourhood radius, which decreases linearly to zero during the self-organizing process. The conventional Euclidian distance is used to determine the best-matching unit (so called 'winner') $\{W(iw, jw)\}$ on a map for the input vector $\{X\}$. Kohonen's SOMs are a type of unsupervised learning. The goal is to discover some underlying structure of the data. Kohonen's SOM is called a topology-preserving map because there is a topological structure imposed on the nodes in the network. A topological map is simply a mapping that preserves neighbourhood relations. In the nets we have studied so far, we have ignored the geometrical arrangements of output nodes. Each node in a given layer has been identical in that each is connected with all of the nodes in the upper and/or lower layer. In the brain, neurons tend to cluster in groups. The connections within the group are much greater than the connections with the neurons outside of the group. Kohonen's network tries to mimic this in a simple way. The algorithm for SOM can be summarized as follows (See Fig.1):

- Assume output nodes are connected in an array (usually 1 or 2 dimensional)
- Assume that the network is fully connected (i.e. all nodes in the input layer are connected to all nodes in the output layer). Use the competitive learning algorithm as follows:
- Randomly choose an input vector x
- Determine the "winning" output node i, where W_i is the weight vector connecting the inputs to output node i. Note the above equation is equivalent to W_i x $\geq W_k$ x only if the weights are normalized.

$$|W_i - X| \leq |W_k - X| \dots\dots\dots\dots\forall k$$

- Given the winning node i, the weight update is

$$W_k(new) = W_k(old) + X(i,k) \times (X - W_k)$$

Where $X(i,k)$ is called the neighborhood function that has value 1 when i=k and falls off with the distance $|r_k - r_i|$ between units i and k in the output array. Thus, units close to the winner as well as the winner itself, have their weights updated appreciably. Weights associated with far away output nodes do not change significantly. It is here that the topological information is supplied. Nearby units receive similar updates and thus end up responding to nearby input patterns. The above rule drags the weight vector W_i and the weights of nearby units towards the input x.

Example of the neighbourhood function is given by the following relation

$$X(i,k) = e^{\left(-|r_k - r_i|^2\right)/(\sigma)^2}$$

Where $\sigma 2$ is the width parameter that can gradually be decreased as a function of time.

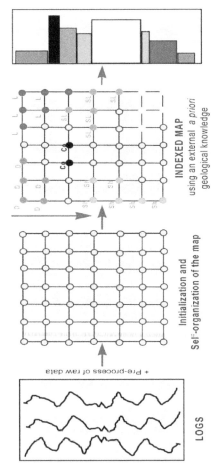

Figure 1. Schematic illustration of the Kohonen's Self-Organizing Map principle

4. The processing algorithm

In this section we train five self-organizing map neural network machines, the inputs of these maps are:

- Data Set1: The five raw well-logs data which are: The Gamma ray, Density, Neutron porosity, Photoelectric absorption coefficient and sonic well- log .
- Data Set2: The estimated Hölder exponents using the continuous wavelet transform of the data set1.
- Data Set3: Data set1 and the three radioactive elements concentrations.
- Data Set4: The estimated Hölder exponents of the data set1 and the Hölder exponents of the radioactive elements concentrations.
- Data Set5: The estimated Hölder exponents of the data set1 and the three radioactive elements concentrations logs.

The goal is to choose the best map that will give more details about lithology of two boreholes named Well01 and Well02 located in the Algerian Sahara.

5. Application on real data

5.1. Geological setting

The Hassi Messaoud field is located in the central part of Algerian Sahara (Figure 2). It is known by its oil-producing wells, mainly from the Cambrian reservoirs. The Hassi Messaoud super-huge field is a structure covering an area of most 1600 km2 and it was discovered in 1956 by well Md1 drilled across the reservoirs in Cambro-Ordovician sandstone at a depth 3337m.The Cambrian deposits which are presented by sandstones and quartzites, are the best known and form the major reservoirs (Cambrian Ri and Ra).

We distinguish in the Cambrian four stratigraphic subdivisions (Algeria Well Evaluation Conference, 2007), which are (Figure 3):

R3: Consisting of 300 m of poorly consolidated microconglomeratic clay sandstones intercalated with clayey siltstone levels that cannot be exploited because of its poor matrix properties and its deep position, below the water table.

R2: Exploitable when in high position, consists of relatively clayey coarse sandstones with intercalated levels of clayey siltstones; the top part of this reservoir, whose thickness is on the order of 40 m, has the best matrix properties.

Ra : the main reservoir, whose thickness varies from 100 m in the east to 130 m in the west, it consists of two major superimposed units which are :

- The lower Ra: with 70 to 95 m as thickness, consisting of medium to coarse sandstones with inter-bedded siltstone levels.
- The upper Ra, which consists of 40 to 60 m of relatively fine clayey sandstones containing skolithos, with many siltstone levels.

4) Ri: Which has 45 to 50 m as thickness and consists of 3 units, produces from 5 to 10m of fine basal sandstones with abundant skolithos; siltstones predominate in the upper units.

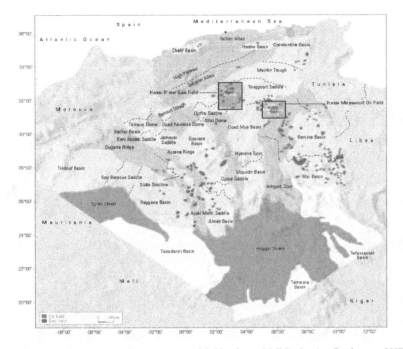

Figure 2. Geographic situation of Hassi Messaoud field (Algeria Well Evaluation Conference, 2007)

5.2. Data description

Well-log is a continuous record of measurement made in borehole respond to variation in some physical properties of rocks through which the bore hole is drilled (Asquith and Krygowski, 2004). In this paper eight well-logs have been processed by the proposed technique of two wells named Well01 and Well02. The exploited well-logging are:

a.   The gamma ray (Gr)

Gamma Ray is a high-energy electromagnetic waves which are emitted by atomic nuclei as a form of radiation. Gamma ray log is measurement of natural radioactivity in formation versus depth. It measures the radiation emitting from naturally occurring Uranium (U), Thorium (Th) and Potassium (K).

b.   The Natural Gamma ray spectroscopy measurements

It measures the total number of Gamma Rays SGR as well as their energy from which is computed the percentage of Potassium (K), Thorium (Th), Uranium (U) and the corrected Gamma Ray from Uranium (CGR)

Total and spectrometry of natural Gamma Ray are also known as shale log. They reflect shale or clay content and used for:

- Correlation between wells.
- Determination of bed boundaries.
- Evaluation of shale content within a formation.
- Mineral analysis.
c. Neutron porosity (Nphi)

The Neutron porosity log is primarily used to evaluate formation porosity, but the fact that it is really just a hydrogen detector should always be kept in mind

The Neutron Log can be summarized as the continuous measurement of the induced radiation produced by the bombardment of that formation with a neutron source contained in the logging tool. which sources emit fast neutrons that are eventually slowed by collisions with hydrogen atoms until they are captured (think of a billiard ball metaphor where the similar size of the particles is a factor). The capture results in the emission of a secondary gamma ray; some tools, especially older ones, detect the capture gamma ray (neutron-gamma log). Other tools detect intermediate (epithermal) neutrons or slow (thermal) neutrons (both referred to as neutron-neutron logs). Modern neutron tools most commonly count thermal neutrons with an He-3 type detector.

The neutron porosity log is used for:

Gas detection in certain situations, exploiting the lower hydrogen density, hydrogen index.

Lithology and mineralogy identification in combination with density and sonic log

d. Density log (Rhob):

The formation density log (RHOB) is a porosity log that measures electron density of a formation. Dense formations absorb many gamma rays, while low-density formations absorb fewer. Thus, high-count rates at the detectors indicate low-density formations, whereas low count rates at the detectors indicate high-density formations. Therefore, scattered gamma rays reaching the detector are an indication of formation Density. The density log is used for:

- Lithology identification combined with neutron and sonic log
- Porosity evaluation
- Gaz beds detection

e. Sonic log (DT):

Acoustic tools measure the speed of sound waves in subsurface formations. While the acoustic log can be used to determine porosity in consolidated formations, it is also valuable in other applications, such as:

- Indicating lithology (using the ratio of compression velocity over shear velocity).
- Determining integrated travel time (an important tool for seismic/wellbore correlation).

- Correlation with other wells.
- Detecting fractures and evaluating secondary porosity.
- Evaluating cement bonds between casing, and formation.
- Determining mechanical properties (in combination with the density log).
- Determining acoustic impedance (in combination with the density log).

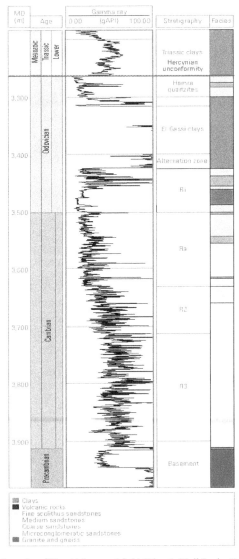

Figure 3. Cambrian stratigraphy of Hassi Messaoud field (Algeria Well Evaluation Conference, 2007)

f.   The photoelectrical absorption coefficient (Pe):

The Photoelectric effect occurs when the incident gamma ray is completely absorbed by the electron. It is a low energy effect hence the photoelectric absorption index, Pe, is measured using the lowest energy window of the density tool.

Pe is related directly to the number of electrons per atom (Z) (Asquith and Krygowski, 2004)

$$Pe = (Z/A)3.6 \qquad\qquad (5)$$

Its unit is barns/electron. It is used also for lithology identification

5.3. Preliminary interpretation of natural gamma ray well-log

Natural gamma radiation occurs in rock formations in varying amounts. Uranium, Thorium, Potassium, and other radioactive minerals are associated with different depositional environments. Clay formations exhibit greater amounts of gamma radiation. A log of gamma radiation will give a positive indication of the type of lithology. Interpretation of gamma log data is done based on the relative low and high count rates associated with respective "clean" and "dirty" environments. Formations having high gamma count rates even though they may exhibit low water saturation are generally unfavorable for production in oil and water well environments

In the description of the Cambrian stratigraphy, this interval is constituted only by sandstones and clays. Thus, our geological interval containing four lithofacies which are: The clay, sandy clays, clayey sandstones and clean sandstones.

This lithofacies classification is based on the gamma ray log value; three thresholds are used to distinguish between these lithologies. We distinguish four lithological units, differed by their gamma ray measurement value, which are:

0<Gr<30Api is a clean sandstone.
30Api<Gr<70Api is a clayey sandstone.
70Api<Gr<90Api is a sandy clay.
Gr>90Api is a clay.

Figures 6a and 7a represent the obtained lithofacies classification based on this approach for the Well01 and Well02 boreholes respectively.

6. Fractal analysis of well-logs data

The first step consists to estimate the Hölder exponents of the eights raw well-logs data of the two boreholes OMJ 842 and WELL02 located in Hassi Messaoud field. The raw well-logs data are: the gamma ray (GR), the Uranium concentration (U), The Thorium concentration Th, the Potassium concentration (K), the slowness (DT), Photoelectric absorption coefficient (Pe), formations density (Rhob) and neutron porosity (Nphi). These data are presented in figures 4 and5.

The Hölder exponents are estimated using the continuous wavelet transform for 929 samples at depths interval [3411.6m-3504.2m] (See figures 4 and 5).

The analyzing wavelet is the Complex Morlet (Morlet et al,1982) defined by :

$$\psi(Z) = \exp(-Z^2 / 2) * \exp(i * \Omega * Z) * (1 - \exp(-\Omega^2 / 4) * \exp(-Z^2 / 2)) \tag{6}$$

Where :

Ω : is the central frequency of the wavelet.

Source codes in C language are developed to calutale the continuous wavelet transform and to estimate the Hölder exponents at each depth.

Ouadfeul and Aliouane (2011) have showed that the optimal value of Ω for a better estimation of the Hölder exponent is equal to 4,8.

Theoretically the Hölder exponent measures the singularity strength. Low exponent indicates a high singularity and a high exponent indicates a low singularity (Audi et al, 2002). Obtained results (figures 4 and 5) show that the main singularities in the raw well-logs data are manifested by spikes in the Hölder exponents graphs.

7. Holder exponents as an input of the Self-Organizing Map

Firstly we have applied the proposed idea at the Well01 borehole, the main depth interval is [3411.6m-3504.2m]. It contains only the four lithological units which are: The Clay, the Sandstone, the clayey sandstone and the sandy clay. The output of the neural machine should be one of these previous lithologies.

For the same reason it is sufficient on this case to use the information provided by the classical interpretation based in the gamma ray log (See figure 6a) for the SOM indexation (Sitao et al, 2003, Gottlib-Zeh et al,1999).

The Numap7.1 software developed by the Neural Networks and Image Processing Lab of Univ. of Texas at Arlington is used for the training and running of the different self-orgnazing maps neural networks.

For each Kohonen's map, the Input is used to train the SOM neural network; in this step weights of connection between neurons are calculated. After that outputs of each map are calculated. Figures 6b, 6c, 6d, 6e and 6f present the output of each Map.

The weights of connection calculated for the Well01 borehole are used to predict lithofacies for Well02, the different type of inputs used for the first well are used for the second one.

In this step we don't need to the Self- Organizing Maps indexations, since the same maps are used. It means that the weights of connections calculated in the training of the first map using the Well01 borehole data and their Hölder exponents are used to calculate the outputs for second well (Well02).

Obtained lithofacies for the Well02 borehole and its corresponding classical interpretation based on the gamma ray log are presented in figure 7.

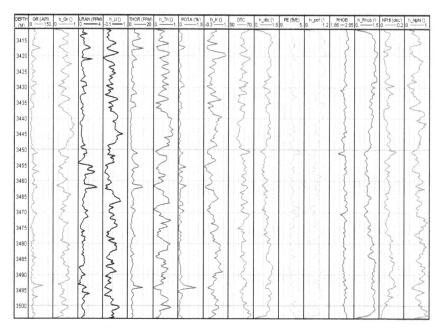

Figure 4. Measured well-logs data for Well01borehole: GR, Vp, RHOB, PEF, NPHI and their corresponding Hölder exponents.

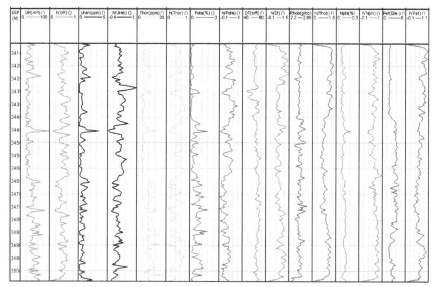

Figure 5. Measured well-logs data for Well02 borehole: GR, Vp, RHOB, PEF, NPHI and their corresponding Hölder exponents.

8. Results discussion and conclusion

By analyzing figures 6 and 7, one can remark that the Self Organizing neural network machines based on the raw well logs data as an input give more details than the classification based on the classical gamma ray interpretation.

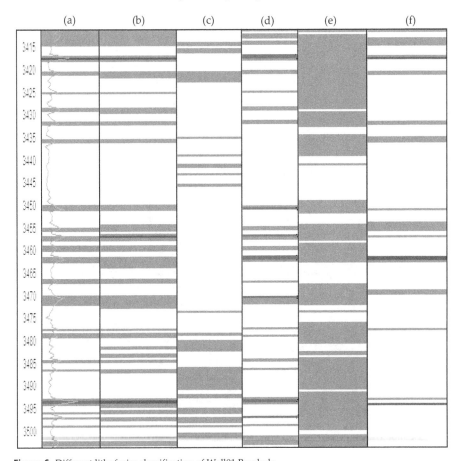

Figure 6. Different lithofacies classification of Well01 Borehole.
(a): Lithofacies classification based on the Gr.
(b): Lithofacies by SOM with data Set1as an input.
(c): Lithofacies by SOM with data set2 as an input.
(d): Lithofacies by SOM with data set3 as an input.
(e): Lithofacies by SOM with data Set4, as an input.
(f): Lithofacies by SOM with data Set 5 an input

Classifications based on the Hölder exponents of the five well-logs data as an input give less details, it means that they can't provide details and thin geological details. However lithofacies prediction based on the five raw well logs data combined with the spectrometric concentration gives more information about shaly character. This is due to the sensitivity of the concentration of radioactive elements to shale.

Finally the Self Organizing map based on the eight raw logs data can give a lot of details and thin facies intercalations. Reservoir model based on the self organizing map neural network machine with the raw data as an input is able to give a detailed information. The self-organizing map neural network model with the Hölder exponents estimated by the continuous wavelet transform as an input is not able to improve the lithofacies classification by the SOM. We suggest by this paper to use always the raw well-logs in a Self-Organizing Map artificial neural network model rather than the fractal analysis using by the CWT, this last processing decrease the details and hide geophysical information that contains the raw data.

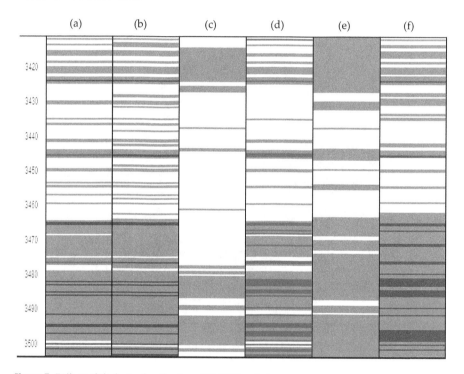

Figure 7. Different lithofacies classification of Well02 borehole.
(a): Lithofacies classification based on the Gr.
(b): Lithofacies by SOM with data Set1as an input.
(c): Lithofacies by SOM with data set2 as an input.
(d): Lithofacies by SOM with data set3 as an input.
(e): Lithofacies by SOM with data Set4, as an input.
(f): Lithofacies by SOM with Data Set 5 an input.

Author details

Leila Aliouane
Geophysics Department, FSTGAT, USTHB, Algeria
Geophysics Department, LABOPHYT, FHC, UMBB, Algeria

Sid-Ali Ouadfeul
Geosciences and Mines, Algerian Petroleum Institute, IAP, Algeria
Geophysics Department, FSTGAT, USTHB, Algeria

Amar Boudella
Geophysics Department, FSTGAT, USTHB, Algeria

9. References

[1] Arneodo, A., Grasseau, G., and Holschneider, M. (1988). Wavelet transform of multifractals, Phys. Rev. Lett. 61:2281-2284.

[2] Arneodo, A., Bacry, E. (1995). Ondelettes, multifractal et turbelance de l'ADN aux croissances cristalines, Diderot editeur arts et sciences.Paris New York, Amsterdam.

[3] Aliouane, L; Ouadfeul, S., Boudella, A., 2011, Fractal analysis based on the continuous wavelet transform and lithofacies classification from well-logs data using the self-organizing map neural network, Arabian Journal of Geosciences, http://dx.doi.org/10.1007/s12517-011-0459-4.

[4] Audit, B., Bacry, E., Muzy, J-F. and Arneodo, A. (2002). Wavelet-Based Estimators of Scaling Behavior, IEEE, vol.48, pp. 2938-2954.

[5] Asquith, G. and Krygowski, D, (2004), Basic Well Log Analysis,. 2nd edition. AAPG Methods in Exploration Series, 28, 244pp.

[6] Dolan, S.S., C. Bean, B. Riollet, 1998, The broad-band fractal nature of heterogeneity in the upper crust from petrophysical logs, Geophys. J. Int. 132 489–507.

[7] Herrmann, F.J.(1997). A scaling medium representation, a discussion on well-logs, fractals and waves, Phd thesis Delft University of Technology, Delft, The Netherlands, pp.315.

[8] Holliger., K., (1996).Upper crustal seismic velocity heterogeneity as derived from a variety of P-wave sonic log, Geophys. J. Int. 125, 813–829.

[9] Crossman, A., and Morlet, J., (1985), Decomposition of functions into wavelets of constant shape, and related transforms, in: Streit, L., ed., mathematics and physics, lectures on recents results, World Scientific Publishing, Singapore.

[10] Gottlib-Zeh, S., Briqueu, L., Veillerette, A., (1999). Indexed Self-Organizing Map: a new calibration system for a geological interpretation of logs, in: Proc. IAMG'99, pp. 183–188

[11] Kohonen, (1992).The self organazing Map, Information Sciences Springer Verlag, New York, 30, pp.312

[12] Kohonen, T. (1998). Self Organization and associative memory pringer Series in Information Sciences, 8, 2nd edn (Berlin :Springer)

[13] Kneib, G. (1995). The statistical nature of the upper continental cristalling crust derived from in situ seismic measurements, Geophys. J. Int. 122, 594–616.

[14] Li, C-F.(2003).Rescaled-Range and power spectrum analysis on well-logging data, Geophys. J. Int. 153, 201–212.

[15] Morlet, J., Arens G., Fourgeau E. and Giard, D. (1982). Wave propagation and sampling theory., Geophysics, 47(2), 203-236.

[16] Mandelbrot, B.B. (1982). The fractal geometry of nature. Ed. W. H. Freeman, San Francisco.

[17] Nump7.1, (2004), Image Processing and Neural Networks Lab, University of Texas at Arlington : http://www-ee.uta.edu/EEweb/ip/new_software.html

[18] Ouadfeul, S., Aliouane, L., (2010), Multiscale Analysis of 3d GPR data using the wavelet transfrom, presented in GPR2010, doi 10.1109/ICGPR.2010.5550177.

[19] Ouadfeul, S., (2006), Automatic Lithofacies Segmentation Using the Wavelet Transfrom Modulus Maxima Lines (WTMM) combined with the detrended Fluctuation Analysis (DFA), 17 the International conference and Exhibition of Turkey, Expanded Abstract.

[20] Ouadfeul, S., and Aliouane, L., (2011b), Multifractal Analysis Revisited by the Continuous Wavelet Transform Applied in Lithofacies Segmentation from Well-Logs Data, *International Journal of Applied Physics and Mathematics*, Vol.1, No.1.

[21] Ouadfeul, S.; Aliouane, L.; Boudella, A., (2012), Heterogeneities analysis from Well-logs data using the Generalzed fractal dimensions and continuous wavelet transform, EGU2012, Abstract.

[22] Pilkington, M. & Tudoeschuck, J.P. (1991). Naturaly smooth inversions with a priori information from well logs, Geophysics, 56, 1811-1818.

[23] Shiomi, K., Sato, H., Ohtake, M., (1997). Broad-band power-law spectra of well-log data in Japan, Geophys. J. Int. 130, 57–64.

[24] Turcotte, D.L. (1997). Fractal and Chaos in Geology and Geophysics, Cambridge University Press, Cambridge,

[25] Grossman, A., and Morlet, J., (1985), Decomposition of functions into wavelets of constant shape, and related transforms, in: Streit, L., ed., mathematics and physics, lectures on recents results, World Scientific Publishing, Singapore.

[26] Gottlib-Zeh, S., Briqueu, L., Veillerette, A., (1999). Indexed Self-Organizing Map: a new calibration system for a geological interpretation of logs, in: Proc. IAMG'99, pp. 183–188

[27] Sitao, W., and Tommy W.S.C, (2003), Clustering of the self-organizing map using a clustering validity index based on inter-cluster and intra-cluster density, Pattern Recognition Society, Elsevier Ltd, doi:10.1016/S0031-3203(03)00237-1

[28] Wu, R.S., Zhengyu, X., Li X.P, (1994). Heterogeneity spectrum and scale-anisotropy in the upper crust revealed by the German Continental Deep-Drilling (KTB) Holes, Geophys. Res. Lett. 21, 911–914.

[29] Algeria Well Evaluation Conference, (2007), http://www.slb.com/resources/publications/roc/algeria07.aspx

Permissions

The contributors of this book come from diverse backgrounds, making this book a truly international effort. This book will bring forth new frontiers with its revolutionizing research information and detailed analysis of the nascent developments around the world.

We would like to thank Dr. Sid-Ali Ouadfeul, for lending his expertise to make the book truly unique. He has played a crucial role in the development of this book. Without his invaluable contribution this book wouldn't have been possible. He has made vital efforts to compile up to date information on the varied aspects of this subject to make this book a valuable addition to the collection of many professionals and students.

This book was conceptualized with the vision of imparting up-to-date information and advanced data in this field. To ensure the same, a matchless editorial board was set up. Every individual on the board went through rigorous rounds of assessment to prove their worth. After which they invested a large part of their time researching and compiling the most relevant data for our readers. Conferences and sessions were held from time to time between the editorial board and the contributing authors to present the data in the most comprehensible form. The editorial team has worked tirelessly to provide valuable and valid information to help people across the globe.

Every chapter published in this book has been scrutinized by our experts. Their significance has been extensively debated. The topics covered herein carry significant findings which will fuel the growth of the discipline. They may even be implemented as practical applications or may be referred to as a beginning point for another development. Chapters in this book were first published by InTech; hereby published with permission under the Creative Commons Attribution License or equivalent.

The editorial board has been involved in producing this book since its inception. They have spent rigorous hours researching and exploring the diverse topics which have resulted in the successful publishing of this book. They have passed on their knowledge of decades through this book. To expedite this challenging task, the publisher supported the team at every step. A small team of assistant editors was also appointed to further simplify the editing procedure and attain best results for the readers.

Our editorial team has been hand-picked from every corner of the world. Their multi-ethnicity adds dynamic inputs to the discussions which result in innovative outcomes. These outcomes are then further discussed with the researchers and contributors who give their valuable feedback and opinion regarding the same. The feedback is then collaborated with the researches and they are edited in a comprehensive manner to aid the understanding of the subject.

Apart from the editorial board, the designing team has also invested a significant amount of their time in understanding the subject and creating the most relevant covers. They scrutinized every image to scout for the most suitable representation of the subject and create an appropriate cover for the book.

The publishing team has been involved in this book since its early stages. They were actively engaged in every process, be it collecting the data, connecting with the contributors or procuring relevant information. The team has been an ardent support to the editorial, designing and production team. Their endless efforts to recruit the best for this project, has resulted in the accomplishment of this book. They are a veteran in the field of academics and their pool of knowledge is as vast as their experience in printing. Their expertise and guidance has proved useful at every step. Their uncompromising quality standards have made this book an exceptional effort. Their encouragement from time to time has been an inspiration for everyone.

The publisher and the editorial board hope that this book will prove to be a valuable piece of knowledge for researchers, students, practitioners and scholars across the globe.

List of Contributors

Reik V. Donner
Research Domain IV – Transdisciplinary Concepts and Methods, Potsdam Institute for Climate Impact Research, Potsdam, Germany

Sid-Ali Ouadfeul
Geosciences and Mines, Algerian Petroleum Institute, IAP, Algeria
Geophysics Department, FSTGAT, USTHB, Algeria

Mohamed Hamoudi
Geophysics Department, FSTGAT, USTHB, Algeria

Manuel Tijera
Department of Applied Mathematics (Biomathematics), University Complutense of Madrid, Spain

Gregorio Maqueda and José L. Cano
Department of Earth Physics, Astronomy and Astrophysics II., University Complutense of Madrid, Spain

Carlos Yagüe
Department of Geophysics and Meteorology, University Complutense of Madrid, Spain

Noboru Tanizuka
Complex Systems Laboratory, Tondabayashi, Japan

O.A. Khachay
Institute of Geophysics UB RAS, Yekaterinburg, Russia

A.Yu. Khachay and O.Yu. Khachay
Ural Federal University, Institute of Mathematics and Computer Sciences, Yekaterinburg, Russia

Asim Biswas and Hamish P. Cresswell
CSIRO Land and Water, Canberra, ACT, Australia

Bing C. Si
Department of Soil Science, University of Saskatchewan, Saskatchewan, Canada

Sid-Ali Ouadfeul
Algerian Petroleum Institute, IAP, Algeria
Geophysics Department, FSTGAT, USTHB, Algeria

Leila Aliouane
Geophysics Department, FSTGAT, USTHB, Algeria
Geophysics Department, LABOPHYT, FHC, UMBB, Algeria

Amar Boudella
Geophysics Department, LABOPHYT, FHC, UMBB, Algeria
Geophysics Department, FSTGAT, USTHB, Algeria

Sid-Ali Ouadfeul
Geosciences and Mines, Algerian Petroleum Institute, IAP, Algeria
Geophysics Department, FSTGAT, USTHB, Algeria

.

Printed in the USA
CPSIA information can be obtained
at www.ICGtesting.com
JSHW011352221024
72173JS00003B/257